An Introduction to

The Practical Study of Crystals, Minerals, and Rocks

K. G. Cox, N. B. Price, and B. Harte

Grant Institute of Geology, University of Edinburgh

McGRAW-HILL Publishing Company Limited

LONDON · New York · Toronto · Sydney

Published by

McGRAW-HILL Publishing Company Limited
MAIDENHEAD · BERKSHIRE · ENGLAND

94053

PRINTED AND BOUND IN GREAT BRITAIN

Preface

This book is based mainly on the first-year course currently taught in the University of Edinburgh. For the purpose of the book, the scope of the course has been slightly enlarged to cater for the needs of departments where emphasis may differ from that which we give in our own laboratories. Our main aim has been to provide the introductory student with a concise companion to his practical work, because, for some time, we have felt that the relevant material was too widely scattered through a number of existing texts. At the same time, we have tried to introduce a realistic appreciation of the student's problems in understanding a subject which is, in most cases, quite unfamiliar to him at the outset. The book, however, is by no means intended to be complete in itself. We assume that the user will concurrently be attending theory lectures or reading theoretical books in geology, and that he has access to adequate practical materials. For this reason, we have not thought it necessary to illustrate the section on mineralogy profusely, because there is no substitute for real minerals. Similarly, aspects of rocks such as distribution and field occurrences are included only in so far as they contribute directly to the understanding of the laboratory course. Descriptions of specific field areas are omitted.

We make no apology for the relatively large amount of space devoted to morphological crystallography. The stereographic approach to crystallography is still one of the most useful training grounds for three-dimensional thinking, as well as being an essential foundation for more advanced mineralogy courses. Experience shows that, for the teaching of this subject to be successful, considerable attention to apparently elementary details is required. This part of the book is designed to be used in conjunction with contact goniometers, crystal models, and much tracing paper.

The section on mineralogy follows the standard form. The number of minerals covered is rather larger than many introductory courses would cover, but we hope this will introduce some flexibility into the use of the book.

In the petrological sections, we have attempted to keep the purely practical aspects of rock description and identification within the capabilities of the introductory student. In the chapter on sedimentary rocks, there is little difficulty in considering genetic features within this framework. Practical study of some of the more important genetic aspects of the igneous and metamorphic rocks is not within the scope of the introductory student, and separate genetic sections have been given at the end of the respective chapters. The inclusion of these sections is justified on the grounds that they are intended to give the student a fuller appreciation of the purpose of careful rock description.

Some anomalies inevitably arise throughout the work, where, for the sake of completeness, we refer occasionally to rocks and minerals which the student is unlikely to be able to identify. We trust, however, that this will not cause any difficulty because these specimens will rarely be available in introductory collections.

K. G. Cox
N. B. Price
B. Harte

Contents

1. Crystallography—Introduction

Crystallography, the study of crystals and the crystalline state, forms a natural introduction to the study of rocks and minerals, since they are predominantly crystalline materials.

Matter is referred to as *crystalline* when the atoms which compose it have a regular arrangement in space, known as the crystal lattice.

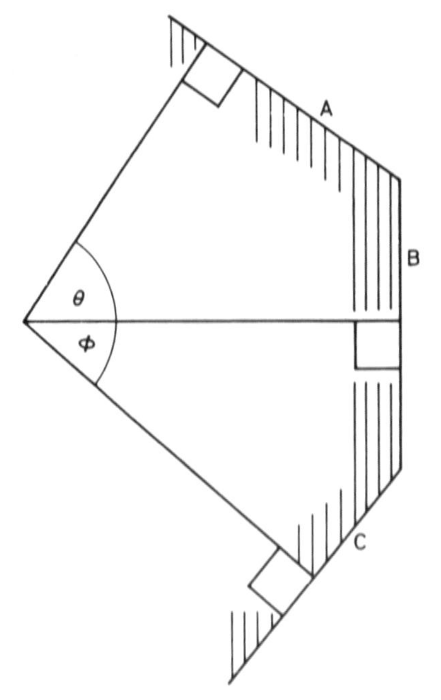

1.1 Interfacial angles. The interfacial angles are A $\wedge$ B $= \theta$, B $\wedge$ C $= \phi$, and A $\wedge$ C $= \theta + \phi$.

The regularity of the internal arrangement is often reflected in an external regularity of form, and the term *crystal* is usually used for fragments of matter showing this feature, though it may also be used for any crystalline fragment providing it is not an aggregate of differently oriented lattices. In this book we shall be largely concerned with the external forms of crystals, a study which is of great importance in the identification of the naturally occurring crystalline substances known as minerals.

Crystals are usually bounded by flat (plane) surfaces termed *faces*. The line formed by the intersection of two adjacent faces is termed an *edge* and the point formed by the intersection of three or more faces is called a *coign* or *solid angle*. The *interfacial angle* between two faces is defined as the angle between the normals to the two faces (see Fig. 1.1). It follows that it is perfectly permissible to speak of the interfacial angle between two faces which are not adjacent.

One of the most important crystallographic concepts is that of the *zone*. This is defined as a set of faces all of which are parallel to a given direction termed the *zone axis* (see Fig. 1.2). Faces in a zone need not be adjacent, but when they are, the edge formed by two such faces is parallel to, or gives the direction of, the zone axis.

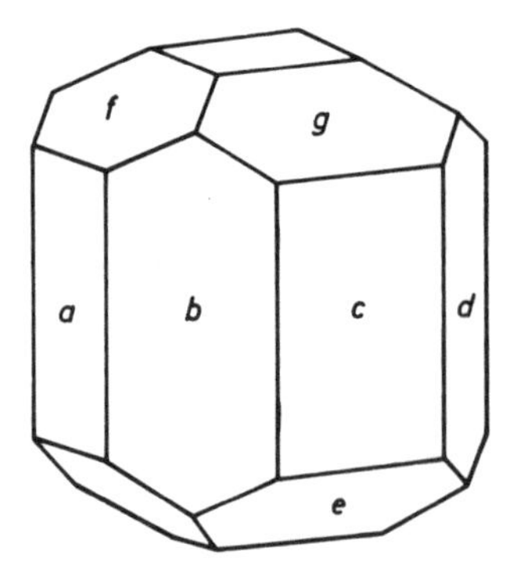

1.2 A crystal of idocrase showing a prominent zone consisting of faces *a*, *b*, *c*, and *d*. Faces *f*, *b*, and *e* belong to a diagonal zone.

It is also useful to distinguish *like* faces and *unlike* faces. Two faces are like if they have the same size and shape in a regularly developed crystal. In Fig. 1.3 the faces marked *a* are all like, as are the faces marked *b*. Faces *a* and *b*, however, are unlike each other. Sets of like faces all have the same relationships to the internal structure (crystal lattice). A complete set of like faces is referred to as a *form* and thus we may speak of the crystal illustrated in Fig. 1.3 as being a combination of two forms.

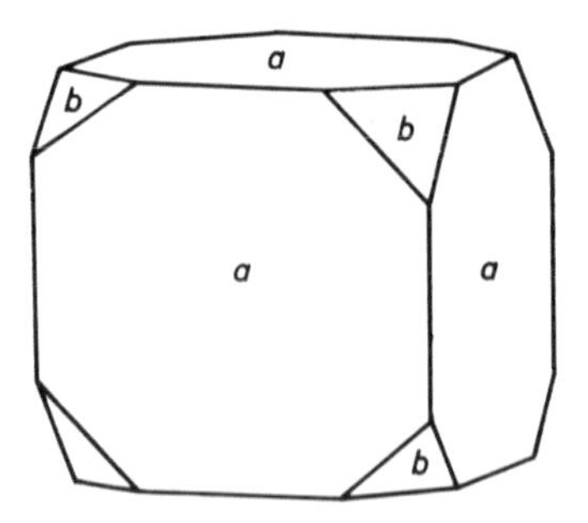

1.3 Like and unlike faces.

Symmetry

Symmetry is the most important of all properties in the identification of crystalline substances. We shall be concerned with the symmetrical arrangement of crystal faces, an arrangement which reflects the internal symmetry of the lattice.

In an initial discussion of symmetry it is convenient to consider *regular* crystals, that is to say crystals in which all the like faces are developed equally and which, therefore, show a symmetry of shape. In a later section we shall refer to the more common situation, in which a crystal is irregularly developed and the shape may superficially appear to be lacking in symmetry.

Symmetry may be described by reference to symmetry planes, axes, and the centre of symmetry. We shall be concerned only with the three simpler symmetry elements described below. A consideration of those is sufficient to meet the needs of the mineralogist dealing with the routine identification of common minerals. A fuller treatment of symmetry, including additional types of symmetry elements is given in standard crystallography text-books. A *symmetry plane* may be considered as a plane along which the crystal may be cut into two equal halves which are mirror images of each other. Figure 1.4 shows the nine symmetry planes of a cube as an example.

A *symmetry axis* is a line about which the crystal may be rotated so that the crystal assumes a position of congruence (i.e., to an observer in a fixed position the crystal presents the same appearance

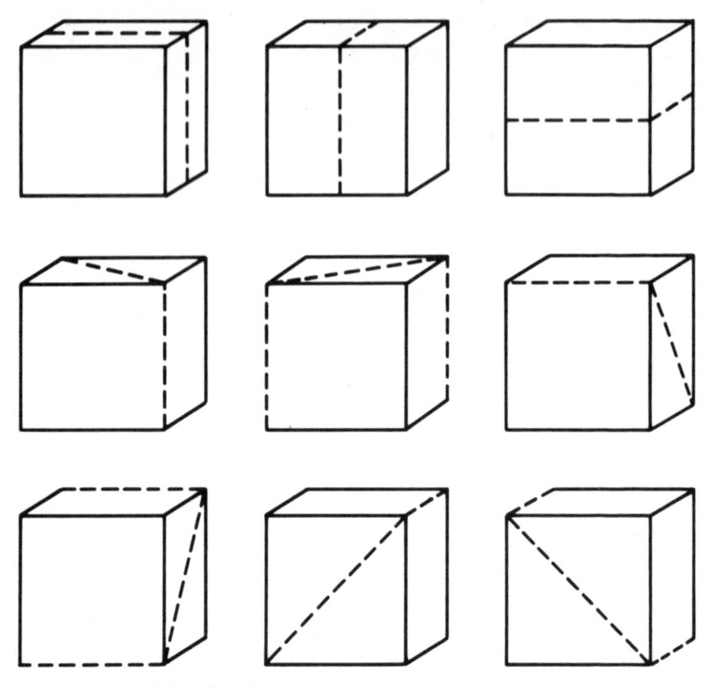

1.4 The nine symmetry planes of the cube.

as it did before rotation commenced) at some stage during the rotation. If a position of congruence occurs after every 180 degrees of rotation the axis is said to be a *diad* or *two-fold* symmetry axis. Other axes may be called *triad*, *tetrad*, or *hexad* (three-fold, four-fold, or six-fold) axes, depending whether congruence is attained every 120, 90, or 60 degrees.

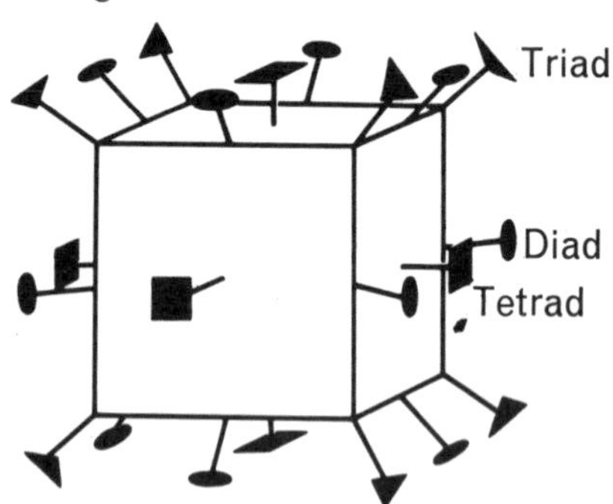

1.5 The thirteen symmetry axes of the cube.

Symmetry axes for a cube are shown in Fig. 1.5. Note also the symbols used to denote axes in diagrams.

A *centre of symmetry* is present when every face of the crystal is matched by one parallel to it on the other side of the crystal. A cube obviously possesses a centre of symmetry, whereas the regular four-faced figure (the tetrahedron) of Fig. 1.6, obviously does not.

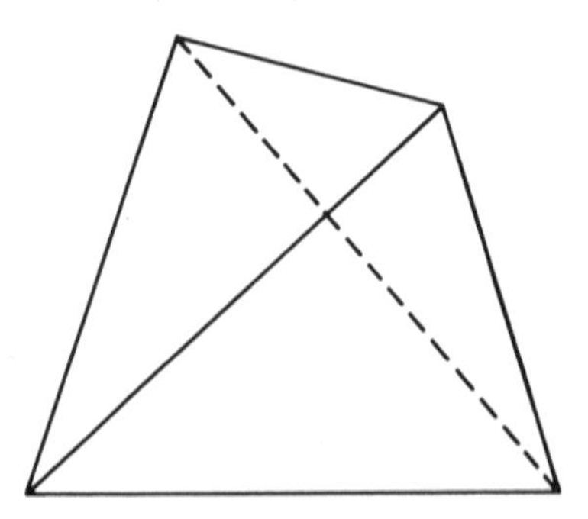

1.6 The tetrahedron, a crystal showing no centre of symmetry. It consists of four faces each of which is an equilateral triangle. Each face is opposite a coign on the other side of the crystal.

The Seven Crystal Systems

On the basis of the symmetry axes present*, crystals can be divided into seven main groups known as crystal systems. These are:

cubic	four triad axes
tetragonal	one tetrad axis
hexagonal	one hexad axis
trigonal	one triad axis
orthorhombic	three diad axes
monoclinic	one diad axis
triclinic	no axes

The axes listed here are *essential* axes of symmetry and many crystals of the cubic, tetragonal, hexagonal, and trigonal systems possess additional axes, e.g., Fig. 1.5 showed that the cube, which is a representative of the cubic system, possesses three tetrad axes and six diad axes in addition to the four triad axes which are the essential feature of the system. Depending on the subsidiary elements of

* Certain groups of crystals are characterized by *inversion axes* rather than the rotation axes we have referred to. The mineral chalcopyrite illustrates this feature and is discussed on p. 57.

symmetry, crystals may be subdivided into thirty-two crystal classes which will not, however, concern us in detail. It is, nevertheless, important to note that *one* class in each system possesses a maximum number of symmetry elements for that system and is known as a *holosymmetric* class. The cube itself, for example, with thirteen axes and nine symmetry planes is a member of the holosymmetric class of the cubic system. The tetrahedron of Fig. 1.6 is also a cubic crystal (i.e., a member of the cubic system) but belongs to one of the lower symmetry classes; e.g., it clearly does not possess any tetrad axes, but it does possess the four triad axes which characterize this system.

Crystallographic Axes

The morphology of crystals can be usefully treated as a type of analytical geometry in three dimensions and it is therefore necessary to have axes to which planes (i.e., faces) may be referred. These reference axes are known as *crystallographic axes* and should not be confused with symmetry axes although they coincide with them to some extent. The crystallographic axes conventionally chosen differ from system to system, this being a matter of convenience in the subsequent mathematical handling of the faces.

Cubic System. We will take the cube itself as an example of a cubic crystal. Clearly the faces of a cube may most conveniently be referred to three axes in space, all mutually perpendicular and parallel to the edge directions of the cube (see Fig. 1.7). Because of the high symmetry of the cube the three axes are indistinguishable. We shall therefore refer to them all by the letter a and because on occasions it is useful to label them more specifically we shall call them a_1, a_2, and a_3.

Tetragonal System. If we imagine a cube stretched or compressed along one of its crystallographic axes the resultant figure has only two square faces left and the other four have become rectangular. This figure may be taken as an example of a holosymmetric tetragonal crystal and has three crystallographic axes, still at right angles, as in the cube, but no longer all identical (see Fig. 1.7). One of the axes, which we shall designate as c, is co-incident with the main symmetry axis of the crystal, the solitary tetrad. The other two axes are still like each other and are called a_1 and a_2 as in the cubic system.

Hexagonal and Trigonal Systems. Ideal hexagonal and trigonal crystals are shown in Fig. 1.7. For geometrical convenience an extra axis is employed in each case so that both systems have a c-axis, coincident with the main hexad or triad symmetry axis, at right angles to a plane containing three identical a- axes set mutually at 60 degrees (a_1, a_2, and a_3).

Orthorhombic System. A figure formed by the 'stretching' or 'compressing' of the cube along *two* of its crystallographic axes may be taken to represent an ideal orthorhombic crystal. Figure 1.7 shows the resultant arrangement of three unequal crystallographic axes, designated a, b, and c, all mutually at right angles.

Monoclinic System. We may imagine a monoclinic crystal to be similar to an orthorhombic crystal but having the a-crystallographic

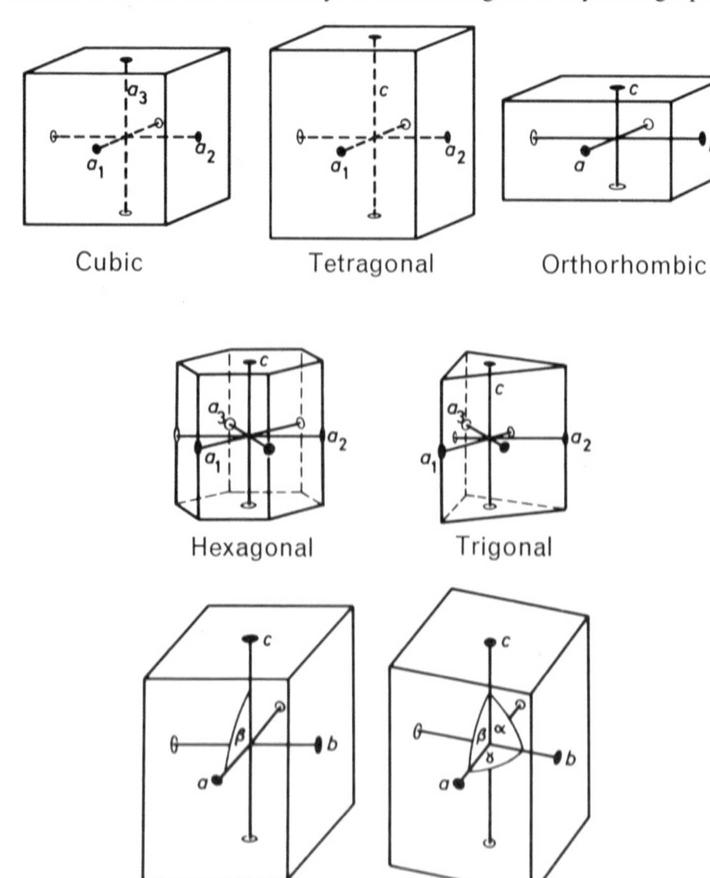

1.7 Crystallographic axes in the seven crystal systems.

axis tilted so that it remains normal to b but not normal to c. Inspection of Fig. 1.7 will show that b remains as a solitary diad axis. The amount of tilting of a relative to c varies from one crystal species to another and the obtuse angle between these two axes is referred to as β.

Triclinic System. In the triclinic system we have the ultimate loss of symmetry axes, and the crystallographic axes, a, b, and c, are all unequal and mutually non-perpendicular. The angular relations of the axes are fully expressed by the angles:

$$a \wedge c = \beta$$
$$a \wedge b = \gamma$$
$$c \wedge b = \alpha$$

Constancy of Angles

So far we have been concerned with regularly developed crystals whose symmetry is evident in their shape. Natural crystals are, however, rarely so perfect; their shapes are distorted and their symmetry is disguised. The *angular* relationships of the faces, however, remain constant, irrespective of the degree of irregularity. We must now adjust our viewpoint of symmetry as being not concerned with sizes and shapes but only with angles. This may be simply illustrated by reference to a crystal of nepheline, a mineral which crystallizes in the hexagonal system. The crystals normally consist of two six-sided faces at either end of a short column made up of six rectangular faces (see Fig. 1.8). Whatever the overall shape of the

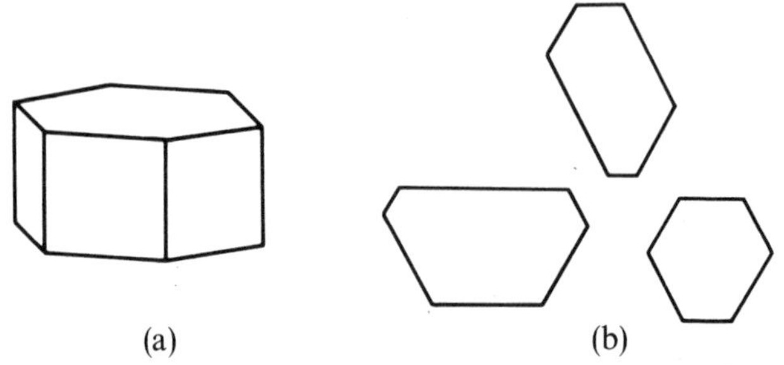

1.8 Constancy of angles. (a) Perspective view of regular nepheline crystal. (b) Shapes of terminal six-sided faces in irregularly developed crystals.

crystals, the interfacial angle between adjacent rectangular faces is always 60 degrees and the angle between any rectangular face and either six-sided face is always 90 degrees. The figure shows variously shaped nepheline crystals exhibiting this constancy of angle.

The importance of the concept of angular symmetry as opposed to symmetry of shape cannot be overstressed, and with a little practice, angular symmetry may be discerned in irregular crystals almost as readily as symmetry of shape is seen in regular crystals. The point may perhaps be made even more forcibly by considering the case of a crystal bounded by six rectangular faces (of three different sorts) with interfacial angles, between adjacent faces, of 90 degrees throughout. This is the shape of the *regular* orthorhombic crystal in Fig. 1.7. *Without further information* it is impossible to say whether this crystal belongs to the cubic, tetragonal, or orthorhombic systems. The angular relations of the faces fit all three cases, and we have no justification in taking the shape of the crystal as significant. The presence of other faces will, however, resolve this problem and it is therefore necessary to discuss the general way of handling all possible faces in more complex crystals.

Miller Indices

The system of face-indexing using Miller indices enables any face to be referred to the crystallographic axes and is somewhat analogous to the giving of co-ordinates to define the position of a plane in analytical geometry. In the simple crystals so far discussed, all the faces have been parallel to at least one, and in many cases to two of the crystallographic axes. Let us now consider the more general situation of a face which cuts all three of the axes, a case such as the face A (the plane XYZ) in Fig. 1.9. The arguments presented here will apply directly to all crystals belonging to systems having three crystallographic axes, i.e., cubic, tetragonal, orthorhombic, monoclinic, and triclinic crystals. Similar treatment can be given for the

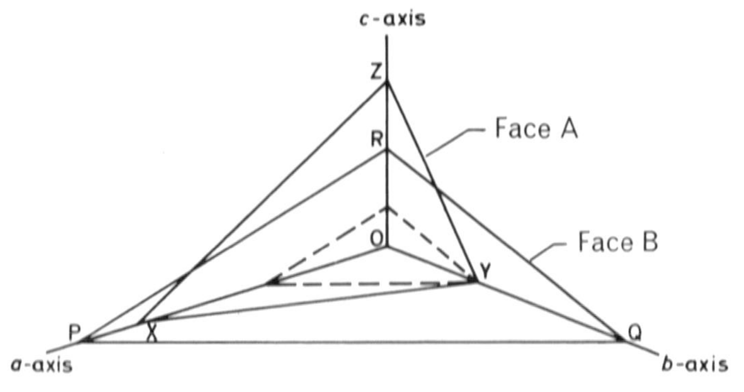

1.9 Illustration of intercepts of faces on crystallographic axes. For explanation see text.

hexagonal and trigonal systems, which have four axes, but will not be included in this work.

The face A has certain angular relationships with the crystallographic axes, which can be determined by the procedure outlined in chapter 3. For the moment it is sufficient to note that this information enables us to calculate the *intercepts* of the face on the three axes. To be more precise, we may determine the *relative* values of the intercepts on the different axes rather than their absolute magnitudes, since the absolute distance of the face from the origin of the crystallographic axes is indeterminate, the crystallographic axes being *directions* having no fixed position.

The ratios of the intercepts on the three axes are expressed in the following form:

$$\frac{OX}{OY} : \frac{OY}{OY} : \frac{OZ}{OY}$$

where OX, OY, and OZ are the intercepts of the *a*-, *b*-, and *c*-crystallographic axes respectively. This arrangement results in the intercept ratio for the *b*-axis always being unity.

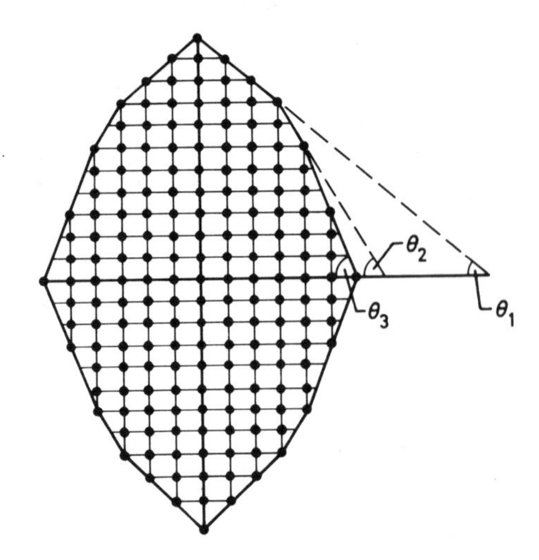

1.10. Rational relationships of parameters. For explanation see text.

It should be clear that in these ratios we have obtained a means of expressing uniquely the angular relationship between the face and the crystallographic axes. In this form, however, the ratios are difficult to handle because they are in general quite *irrational* numbers. In order to avoid this difficulty it is necessary to consider several other faces on the same crystal. Rationality can be introduced by this method, which depends on the fact that although the intercept ratios (generally termed the *parameters*) of any one face are irrational, the parameters of any two faces are always *rationally* related to each other.

An example should make this point clear. Crystals, as we have seen, have an internal structure which consists of regularly repeating units. Crystallographic planes such as faces must clearly have some sort of rational relationship to these units. This is illustrated in Fig. 1.10 where a representation of a crystal built on a repeating rectangular unit of pattern is given. The crystal is bounded by three different types of face, each of which may be imagined to intersect the two crystallographic axes lying in the plane of the diagram. The angles between each of the three faces and the horizontal crystallographic axis are represented by θ_1, θ_2, and θ_3. Thus the intercept on the vertical axis is related to the intercept on the horizontal axis by the following expression:

$$\frac{\text{Intercept on vertical axis}}{\text{Intercept on horizontal axis}} = \tan\theta$$

Hence, if we equate the horizontal axis with the *b*-crystallographic axis and regard the intercept of any face on this axis as unity, we can extend this relationship into the form:

$$\text{Parameter of face} = \tan\theta$$

Thus for the three faces represented on the diagram we have the three parameters $\tan\theta_1$, $\tan\theta_2$, and $\tan\theta_3$, respectively. Each of these values will be in general quite irrational, an irrationality which will depend on the dimensions of the unit of pattern. Inspection of the figure will show, however, that the following relationships exist, namely:

$$\tan\theta_2 = 2\tan\theta_1 \quad \text{and} \quad \tan\theta_3 = 3\tan\theta_1$$

Thus the parameters are related to each other in a rational and also in a simple way. This general argument can be extended to an infinite number of faces at different angles, because the relationship:

$$\text{Parameter of the face} = \frac{n_1}{n_2}\tan\theta$$

where n_1 and n_2 are any two integers, is always satisfied. In practice, however, faces with remote numerical relationships to each other are extremely rare whereas simple ratios between parameters are very common.

Returning to Fig. 1.9, let us now consider a second face, B, in relation to the original face, A. Let us suppose that the intercepts of the two faces on the crystallographic axes have been determined as follows:

Face A, OX = 4·20; OY = 1·50; OZ = 2·80
Face B, OP = 5·05; OQ = 3·60; OR = 1·68

The intercepts may now be converted to parameters by dividing each set by the appropriate *b*-axis intercept, thus:

Parameters of Face A	*Parameters of Face B*
4·20/1·50 = 2·80	5·05/3·60 = 1·40
1·50/1·50 = 1·00	3·60/3·60 = 1·00
2·80/1·50 = 1·86	1·68/3·60 = 0·465

It can now be seen that face B has parameters which are simply and rationally related to those of face A. This can be illustrated in Fig. 1.9 by imagining that face B is moved parallel to itself to the position shown by the broken line, so that the intercepts of the two faces on the *b*-axis have become coincident. The intercept of A on the *a*-axis, as the parameter calculation shows, can now be seen to be twice that of the face B, while the intercept of A on the *c*-axis is four times that of B. It is most important to understand that a general result of this type can be obtained for any two faces on the same crystal, irrespective of the crystal system and providing that crystallographic axes have been selected with due regard to the obvious crystallographic directions shown by the crystal, irrespective of the arrangement of the axes.

A simple system of designating fully the angular relationships of faces can now be devised and depends on the selection of *one* face as a standard to which all other faces may be referred. Such a selected face is referred to as the *parametral plane* for the particular substance concerned. The precise conditions governing the choice of a parametral plane will be discussed in chapter 3; for the moment it is sufficient to note only that it must be a face which cuts all three crystallographic axes.

Let us suppose that face A in Fig. 1.9 has been selected as the parametral plane. It is now possible to designate the full angular relationships of face B by means of a simple set of indices, providing that the parameters of the parametral plane itself are also specified.

These parameters are known as the *axial ratios* of the crystal, and in the present case since the parameters of face A have already been calculated we can say that the axial ratios are $a:b:c = 2{\cdot}80:1:1{\cdot}86$. This is what is meant by the 'lengths' of crystallographic axes and enables us to speak loosely of axes being 'longer' or 'shorter' than other axes. The axial ratios of a substance are of course a very diagnostic property, and with the techniques of reflection goniometry may be determined with considerable accuracy. Formerly this constituted one of the most important methods of mineral identification but has now been largely superseded by X-ray methods.

Having defined the parametral plane, face B could be indicated by a symbol such as $\frac{1}{2}$, 1, $\frac{1}{4}$ which would mean that its *a*-axis parameter was half that of the *a*-axis parameter of the parametral plane, its *b*-axis parameter was the same as that of the parametral plane and its *c*-axis parameter was a quarter that of the parametral plane. This system however introduces an inconvenient element, in that the common faces which are parallel to an axis have parameters of infinity for that axis. To avoid the mathematical difficulties which this introduces into subsequent calculations the system due to Miller involves taking the reciprocals of the parameter ratios. Hence the *Miller indices* of face B are (214), the reciprocal of ($\frac{1}{2}$, 1, $\frac{1}{4}$). The round bracket is a convention signifying that the indices refer to a face (other types of bracket will be used in chapter 3 to refer to zone axes and forms). Note that no commas are used between the digits, which refer to the three crystallographic axes in the order *a*, *b*, *c*. In speech this index is given as 'two one four', not 'two hundred and fourteen'.

The steps followed in the derivation of the Miller indices for the face B of Fig. 1.9 are recapitulated in Table 1.1. A few additional features of Miller indices may also be conveniently noted at this stage. Firstly, a set of indices should always be reduced, by division by any possible common factor, to a minimum value for each digit, i.e., the indices (428) represent the same direction in space as the indices (214). The latter set of indices is the correct set to use. Secondly, if a zero appears in the indices the face concerned is parallel to one crystallographic axis, and if two zeros appear the face is parallel to two axes. In such cases the remaining digit will always be unity. Thirdly, the crystallographic axes have positive and negative ends, and an intersection of a face with the negative end of an axis will be represented by a bar above the digit referring to that axis in the set of indices.

Table 1.1

STEPS FOLLOWED IN THE DERIVATION OF THE MILLER INDICES FOR THE FACE B OF FIG. 1.9

	Face A			*Face B*		
	a-axis	*b*-axis	*c*-axis	*a*-axis	*b*-axis	*c*-axis
Intercepts of face as shown in the figure	4·20	1·50	2·80	5·05	3·60	1·68
Parameters of face	2·80	1·00	1·86	1·40	1·00	0·465
Ratio of parameters to those of the parametral plane (taking this to be face A)	1	1	1	$\frac{1}{2}$	1	$\frac{1}{4}$
Reciprocals of the above ratios	1	1	1	2	1	4
Miller indices	(111)			(214)		

Two faces which are parallel to each other on opposite sides of the crystal thus have Miller indices which are numerically identical but opposite in sign, e.g., reverting to the previous example, face B of Fig. 1.9, which was assigned the indices (214) is matched on the

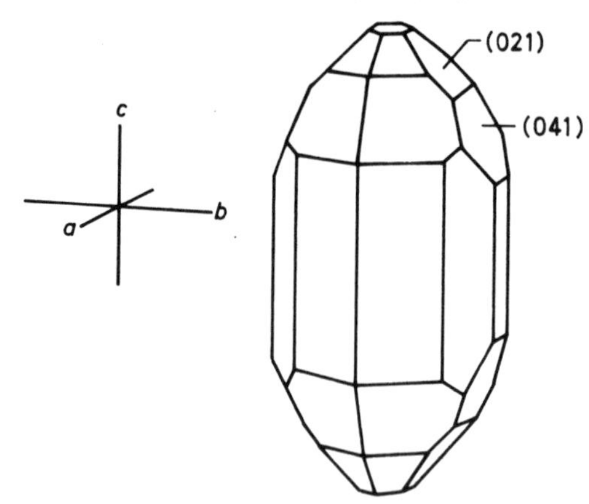

1.11 A crystal of topaz (orthorhombic) illustrating the reciprocal relationship in Miller indices.

opposite side of the crystal by a parallel face ($\bar{2}1\bar{4}$). This face would be referred to in speech as 'bar two, bar one, bar four'.

It is also useful to become thoroughly familiar with the implications of the reciprocal factor in the calculation of the indices. Thus in the simple orthorhombic crystal illustrated in Fig. 1.11, two faces indexed as (021) and (041) are shown. The reciprocal factor means that the second face, which has a *larger* index for the *b*-axis has a *smaller* intercept on that axis.

Miller–Bravais Indices

The Miller indices discussed above are three-digit symbols used in those crystal systems which have three crystallographic axes. In the trigonal and hexagonal systems where four crystallographic axes are employed, four-digit symbols known as Miller–Bravais indices are used. For a face these have the general form ($xyuz$) where x, y, and u refer to the intercepts on the three equal a-axes, and z refers to the intercept on the c-axis. The symbols x, y, and u are not independent of each other because they refer to three co-planar axes at fixed angles to each other. As a result of this the following relationship always holds in Miller–Bravais indices:

$$x + y + u = 0$$

Further reference to this point will be made in chapter 3 in the section concerned with the assigning of indices to faces.

2. The stereographic projection

It is now necessary to discuss the handling of crystallographic data in order to determine the symmetry, types of faces present, and axial ratios of an unknown crystal. These features may all contribute to the identification of a crystalline substance, the essential preliminary objective of the mineralogist. The stereographic projection is the most commonly used system.

Imagine that a crystal is placed at the centre of a sphere (Fig. 2.1) and that the normal to each face (termed the *face-pole*) projects from the centre of the sphere and intersects the surface of the sphere. At this stage the essential angular relationships of the faces have been

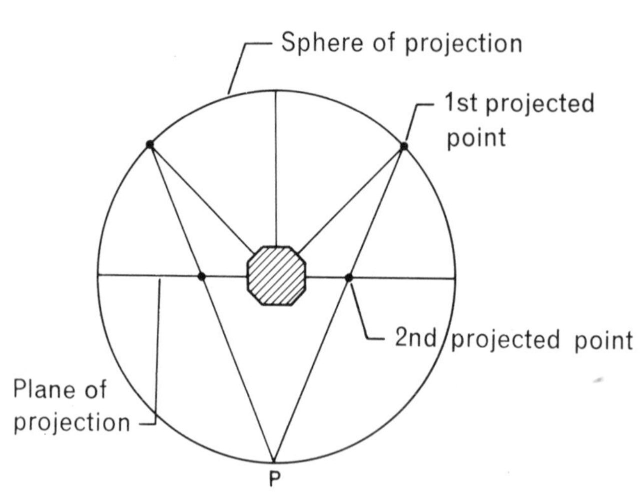

2.1 Principle of the stereographic projection. A section cut through the sphere of projection normal to the equatorial plane (the plane of projection).

transferred to the sphere as a series of points. Providing we imagine the crystal to be very small relative to the sphere, all irregularities of shape are lost and only the angular relationships, which as we have seen, are constant for a given mineral, are preserved.

Each point on the upper hemisphere of the sphere is now projected towards the south pole of the sphere (P) and the line of projection intersects the equatorial plane at a new point. The equatorial plane is termed the plane of projection and is a two-dimensional surface on which the upper half of the crystal, a three-dimensional object, is now represented by a series of points, each one a face-pole. The circle bounding the plane of projection, that is, the original 'equator' of the sphere of projection, is known as the *primitive circle*.

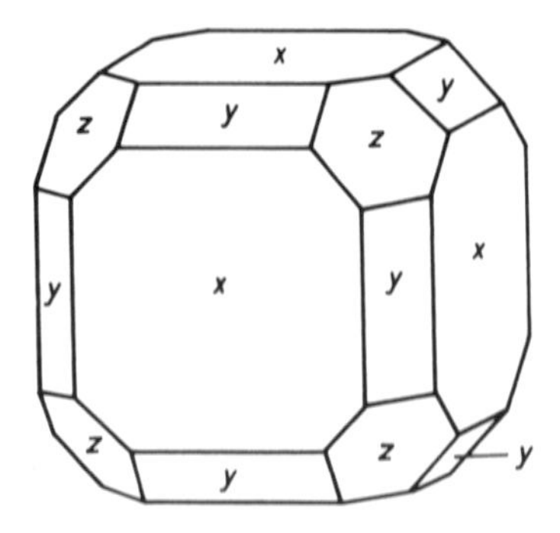

2.2 A crystal belonging to the cubic system showing the three forms: *x* cube, *y* rhombdodecahedron, and *z* octahedron.

A suitable crystal to illustrate the stereographic projection (stereogram) is shown in Fig. 2.2. This crystal belongs to the cubic system and is a combination of several different types of face, e.g., faces marked *x* are the faces of a cube and there are six of them. Faces marked *y* symmetrically bevel the original edges of the cube (twelve in number) and faces marked *z* symmetrically truncate the original eight coigns of the cube. This introduces the idea of *form*. All the *x* faces are like each other and represent one form known as the *cube*. The *y* faces represent the *rhombdodecahedron* form, and the *z* faces the *octahedron* form. This crystal then is a combination of three different forms, each of which can exist independently as shown in Fig. 2.3. The interfacial angles between cube faces and rhombdodecahedron faces are all 45 degrees, whereas the angle between a cube face and an adjacent octahedral face is about 55 degrees and the angle between octahedral faces and rhombdodecahedral faces is about 35 degrees.

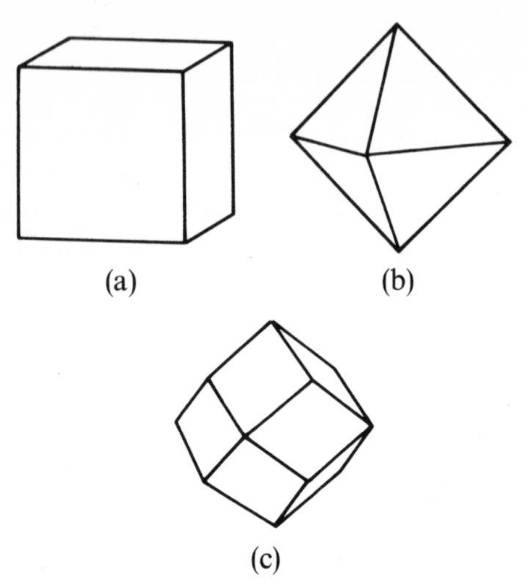

2.3 The three forms shown in Fig. 2.2 developed separately. (a) cube; (b) octahedron; (c) rhombdodecahedron.

A stereogram of the crystal (upper faces only) is shown in Fig. 2.4. It should be noted that every point on the stereogram represents a *direction* in space and the distance between any two points represents the *angle* between two directions. It is important to note, however,

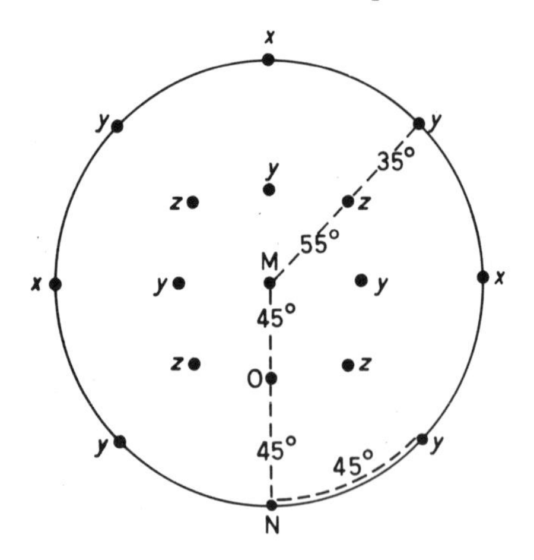

2.4 Stereogram of the combined cube-octahedron-rhombdodecahedron.

that the scale varies in different parts of the stereogram (e.g., face O, one of the rhombdodecahedron faces, is at 45 degrees to faces M and N, the two adjacent cube faces, but does not appear half-way between them on the stereogram).

Zones and Great Circles in the Stereogram

A zone was defined earlier as a set of faces all parallel to a single direction known as the zone axis. In the crystal we have been considering, several zones may be discerned (see Fig. 2.5), some of which are listed below:

(i) *a*, *b*, *c*, *d* parallel to a vertical zone axis marked by the direction of the *bc* edge (or the *cd* edge, etc.).
(ii) *b*, *f*, *g* parallel to zone axis marked by the *bf* edge.
(iii) *d*, *h*, *g* parallel to zone axis marked by *dh* edge.
(iv) *j*, *f*, *k*, *d* a diagonal zone with zone axis parallel to the edge *dk*.

Each of these zones contains further faces out of sight on the other side of the crystal.

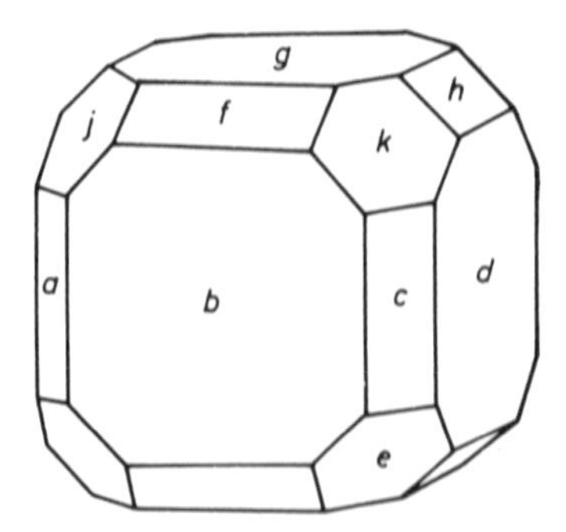

2.5 The combined cube-octahedron-rhombdodecahedron showing zones: e.g., *abcd*, *bfg*, *dhg*, and *jfkd*.

It is clear that the face-poles of all faces in a zone must lie in a plane, the normal to which is the zone axis. It follows that the plane cuts the sphere of projection in a *great circle*, i.e., a circle on the sphere, whose centre coincides with the centre of the sphere. In the stereogram itself great circles project as arcs or straight lines (arcs of infinite radius), which can be identified by their property of cutting the primitive circle at diametrically opposed points. Figure 2.6 gives a complete stereogram of the combined cube–octahedron–rhombdodecahedron crystal with some of the zones marked. In this figure the faces on the lower half of the crystal are shown by open circles. Their position is found by projecting the point where the face-poles cut the lower hemisphere of the sphere of projection *upwards* towards the north pole of the sphere. Since this crystal has a plane of symmetry coincident with the plane of projection it follows that every point, within the primitive, representing an upper face is coincident with the projection of a lower face. Hence each solid point (an upper face) coincides in the stereogram with an open circle

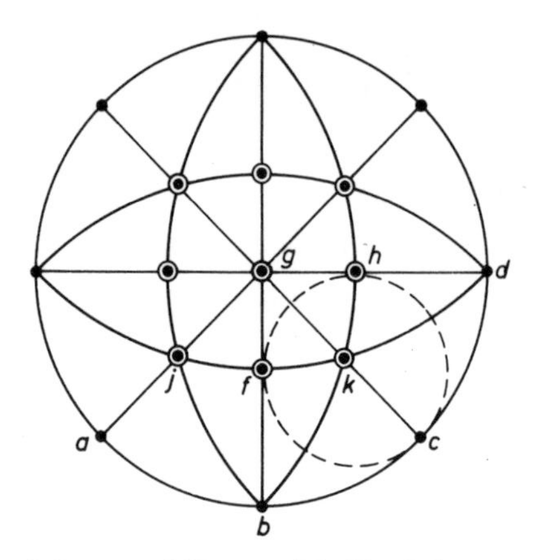

2.6 Stereogram of the crystal illustrated in Fig. 2.5. A small circle centred on *k* has radius 35 degrees and links the three rhombdodecahedron faces *f*, *h*, and *c*, adjacent to *k*.

(a lower face). Faces which project onto the primitive itself lie in the plane of symmetry and are hence not repeated.

Small Circles

A small circle on the surface of a sphere is defined as any circle which is not a great circle. On the surface of the sphere the small circle joins points all of which are equidistant from the centre of the small circle. In terms of angles, the small circle links points which are all an equal angular distance from the centre of the small circle. Small circles on the sphere of projection project as circles into the plane of projection but the stereographic centre no longer coincides with the geometrical centre of the projected circle. As an example of this effect, a small circle (marked by a broken line) has been added to Fig. 2.6 and has the face *k* as its stereographic centre, i.e., all points on the broken line are angularly equidistant from *k*. In this case the radius of the small circle has been chosen as 35 degrees, the interfacial angle between an octahedron face and the three adjacent rhombdodecahedron faces. Thus the small circle passes through the face poles of the faces *f*, *h*, and *c*, the relations of which to the face *k* may be seen in Fig. 2.5. In the latter figure it should be readily appreciated that the three rhombdodecahedron faces are, in terms of angles, equidistant from the octahedron face, *k*. This should be contrasted with the position of the face pole of *k* relative to the other three faces in Fig. 2.6 where, because of the non-linearity of the scale in stereographic projections, *k* does not lie at the geometrical centre of the small circle.

Small circles are particularly important in the stereographic projection because of their use in positioning face-poles when a stereogram is being constructed from a measured crystal. The interfacial angles between two faces already plotted and a third unplotted face-pole are measured. Small circles of the appropriate angular diameter are then plotted centred on the two known poles and their intersection gives the position of the unknown pole.

The Stereographic Net

In order to avoid laborious geometrical constructions a stereographic net, also termed a Wulff net (Fig. 2.7) is employed. The net may readily be visualized as a globe and consists of the following elements:

(a) A family of great circles at two-degree intervals, equivalent in global terms to meridians of longitude. It will be seen that the primitive circle is the only one of the family which appears as a full circle and that the centremost great circle appears as a straight line.

(b) A family of small circles, equivalent to parallels of latitude. The centremost of the family (the 'equator') is in fact a great circle and appears as a straight line.

The stereogram is constructed on tracing paper which is laid over the net and rotated to any desired position. Any required great circle can be drawn through given points and small circles of any desired radius can be drawn round any point on the primitive circle. Small circles centred on points inside the primitive cannot, however, be drawn from the net and must be constructed separately.

The stereographic net is also used as a scale for measuring angles. A great circle course is the shortest distance between two points on the surface of the globe, and by analogy, interfacial angles between two face poles must always be measured along the great circle

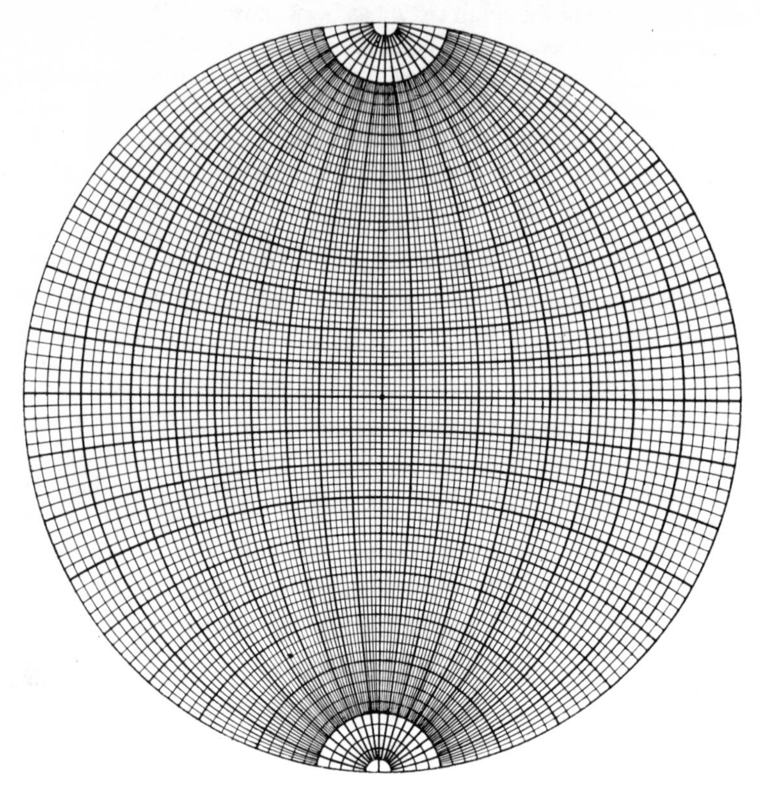

2.7 The stereographic or Wulff net.

connecting the poles, the small circles on the net acting as the scale. Thus, the net is adjusted until the two poles lie on the same great circle and the angle between them is determined by counting the number of small circles crossing the great circle between the two points (see Fig. 2.8).

If it is desired to measure the distance between two points which lie in different hemispheres the arrangement of the net is as shown in Fig. 2.9.

Construction of Small Circles centred inside the Primitive Circle

It is occasionally necessary to construct small circles centred within the primitive circle, since these cannot be drawn with the stereographic net. The principle of this and many other constructions lies in using the primitive circle in a dual role and imagining it also to represent the *sphere of projection* viewed from the side, that is from a viewpoint lying in the plane of projection produced.

As a preliminary, suppose it is required to measure the angle

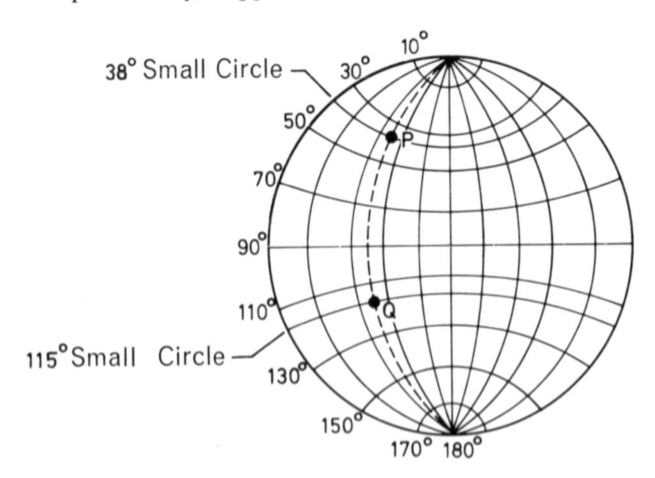

2.8 Measurement of angle between two points, using the net. The stereogram, drawn on tracing paper, is rotated over the net until the two points, P and Q, lie on the same great circle. The angular distance between P and Q is then read using the small circles as scale marks. In this case the angle is 115 − 38 = 77 degrees.

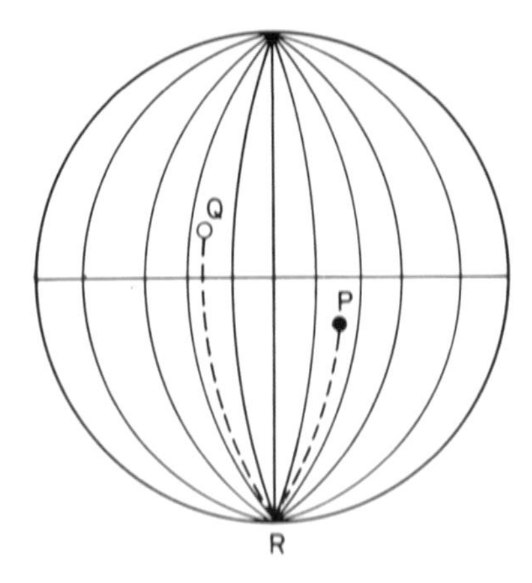

2.9 Measurement of angular distance between two points when one point is on the lower hemisphere. The points P and Q are arranged to lie on symmetrically related great circles as shown. The angular distance PQ = PR + RQ.

between a pole P and the centre of the stereogram, O (see Fig. 2.10). Imagine initially that there is an axis XY lying in the plane of projection so that the poles of P and O project onto it in the stereogram. Then imagine that the crystal is rotated through 90 degrees so that O moves to O′. XY now represents the original plane of projection viewed edge on, while the original primitive circle now represents the sphere of projection viewed from the side. Because of the principle on which stereograms are made, it should now be clear that

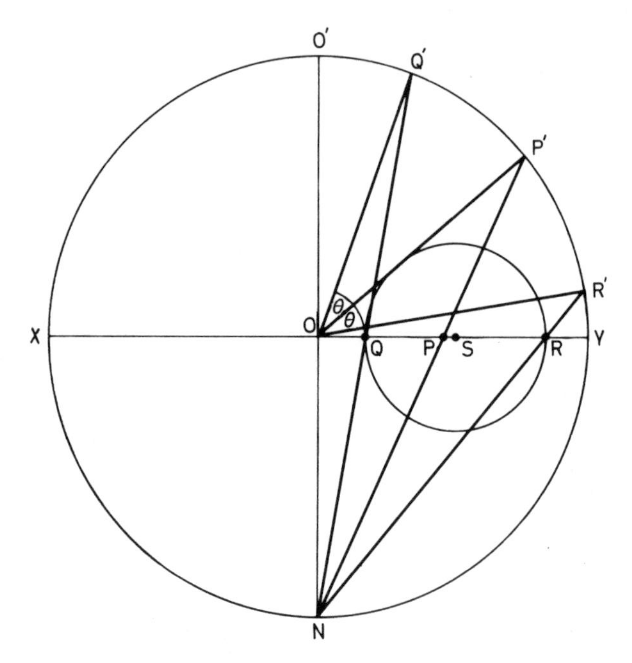

2.10 Construction of small circles. For explanation see text.

there is a point P′ which represents the position of the face-pole P on the surface of the sphere of projection, and it is found by projecting a line from the south pole of the sphere of projection, N, through the point P. The angle between OP′ and OO′ is the angle required, the angle between the face-pole of P and the vertical. It can now be measured with a protractor in the diagram.

The measured angle (O′OP′) is represented on the stereogram by the linear distance OP. This should help to emphasize the non-linearity of the angular scale in stereograms in general since it can now be appreciated that as P moves further away from O in the stereogram, the angle O′OP′ does not change linearly with the change in the distance OP.

The principle of the construction of a small circle of any desired radius about P should now be readily understandable. In Fig. 2.10 by analogy with the previous arguments, the pole P is represented on the sphere of projection by P′, while the points Q′ and R′ are the intersections of two poles, each of which is θ degrees distant from P′, with the sphere of projection. Now one can imagine that a small circle on the surface of the sphere of projection, centred on P′ and with a radius of θ degrees, lies with its plane normal to the paper and passes through the points Q′ and R′. By imagining rotation through 90 degrees so that the plane of projection becomes parallel again with the plane of the paper, the points Q′ and R′ may be represented by Q and R, the stereographic positions of the two poles θ degrees from P. If we accept the fact that a small circle on the sphere of projection remains circular when transferred to the plane of projection, all that remains is to construct a circle through Q and R with diameter QR. The geometrical centre of this circle, S, lies at the midpoint of QR, and again one should notice that it is not co-incident with the stereographic centre, P.

The example chosen here is the simplest case, the method being slightly different when the angle θ is such that part of the small circle lies on the lower hemisphere of the sphere of projection. The student will find it an instructive exercise to work out the necessary construction.

Construction of a Stereogram from a Crystal or Crystal Model

A drawing or drawings of the crystal should first be made and letters or numbers assigned to the different faces (see Fig. 2.11). Prominent zones are then selected and interfacial angles within each

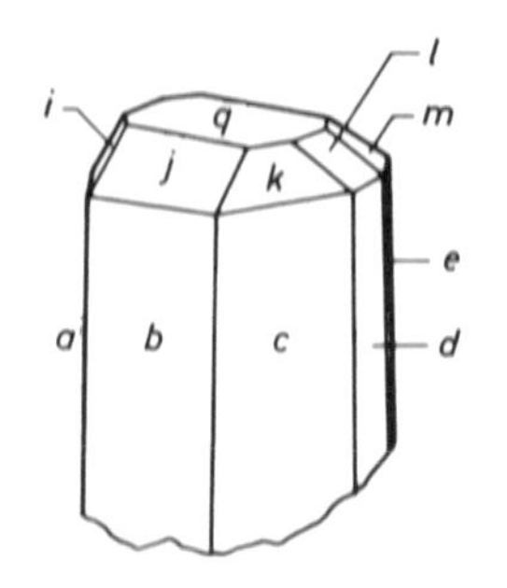

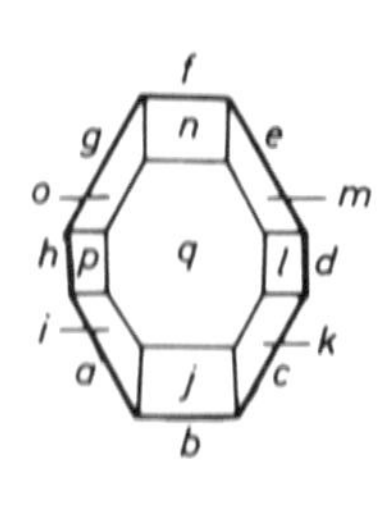

2.11 Drawings of crystal used to illustrate the construction of a stereogram.

zone are measured, most conveniently using a contact goniometer, until every face has been included. The sum of interfacial angles in a complete zone should be 360 degrees. The results are tabulated as shown below:

Zone 1	*Zone 2*	*Zone 3*	*Zone 4*	*Zone 5*
a	*b*	*h*	*c*	*a*
62°	50°	34°	30°	30°
b	*j*	*p*	*k*	*i*
62°	40°	56°	60°	60°
c	*q*	*q*	*q*	*q*
28°	40°	56°	60°	60°
d	*n*	*l*	*o*	*m*
28°	50°	34°	30°	30°
e	*f*	*d*	*g*	*e*
62°				
f				
62°				
g				
28°				
h				
28°				
a				

A prominent zone is then selected and plotted in the primitive circle (see Fig. 2.12). In this case Zone 1 is chosen. The centre of the stereogram is marked; the pole of *a* is positioned arbitrarily and the remaining face-poles marked off around the primitive using the intersections of small circles with the primitive circle of the net as degree marks.

Face *q* may now be positioned since from Zone 2 the angle $b \wedge q$ is seen to be 90 degrees, and from Zone 3 the angle $d \wedge q$ is also 90 degrees. The net is used to draw a 90-degree small circle centred on *b* and then on *d*. In both cases this is a straight line (the 'equatorial' circle) and the two intersect in the centre of the stereogram which is the position of face *q*. The zones *bqf* (Zone 2) and *hqd* (Zone 3) may now be drawn in and are straight lines since the zone axis is in both cases horizontal (i.e., in the plane of projection). Faces *j* and *n* lie in Zone 2 and may now be plotted. The net is turned until the zone coincides with one of the great circles, in this case the central linear great circle, and *j* is positioned on the zone, 50 degrees from *b* and 40 degrees from *q*. Face *n* is found similarly, as are faces *l* and *p* in Zone 3.

Zones 4 and 5 also have zone axes in the plane of projection and are represented by the straight lines *cqg* and *aqe*. Hence the remaining faces *i*, *k*, *m*, and *o* may be positioned.

Further zones which were not apparent previously can now be distinguished. These are the diagonal zones *bklmf*, *bipof*, *hijkd*, and *honmd*; zones such as *apne*, *ajle*, *gpjc*, *gnlc*, and a number of zones including only three faces, such as *cmg*. The completed stereogram is shown in Fig. 2.12.

As a further example of the use of intersecting small circles in plotting face-poles of unlocated faces, consider the case of a small face *r* truncating the coign formed by the intersection of faces *b*, *c*, *j*,

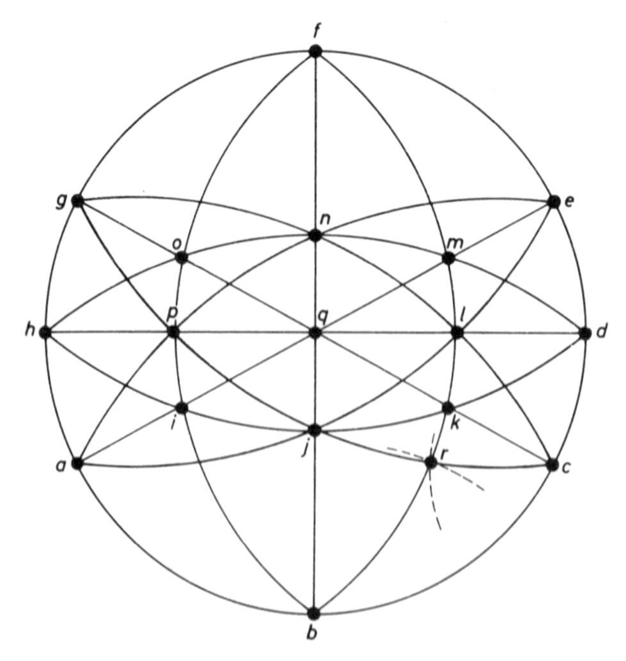

2.12 Stereogram of the crystal illustrated in Fig. 2.11. Dashed arcs are small circles centred on *b* and *c*, used in the location of *r*.

and *k*. Suppose measurement gives $c \wedge r = 31$ degrees and $b \wedge r = 48$ degrees. Small circles centred on *c* and *b* of radii 31 and 48 degrees respectively are drawn using the net. The intersections of these (dashed lines) and the position of *r* are shown in Fig. 2.12. Note that *r* falls on two of the zones already established.

3. Determination of crystal morphology

Determination of Symmetry

Having constructed a stereogram of an unknown crystal by the procedures outlined in the last chapter, the next step in the investigation of the properties of the crystal is the determination of its symmetry elements. This will involve the identification, within the stereogram, of mirror planes and symmetry axes as defined in chapter 1. It is not until this step has been carried out that the most important part of the investigation, the determination of the crystal system, can be made.

Symmetry planes and axes can be identified by two methods, but both require an intelligent appraisal by the investigator as to their possible position, before the necessary tests can be carried out. To make this point clear, consider the stereogram given in Fig. 3.1.

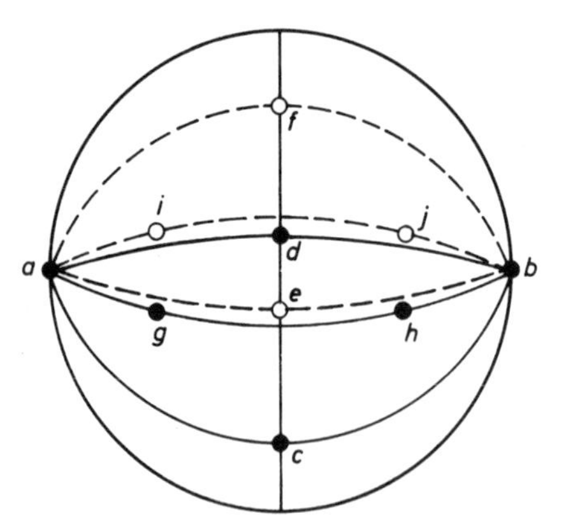

3.1 Stereogram with zones marked by great circles. Portions of great circles lying on the lower hemisphere are shown by broken lines.

One thing is clear immediately from this diagram, that the zone through faces *cdef* is a symmetry plane, that is, it divides the diagram into two halves which are mirror images of each other. It is by no means obvious, however, that there are two other symmetry planes shown by this crystal, that is to say, firstly, the zone *aghbji*, and secondly a plane at right angles to this, not coinciding with a zone. Only practice will enable the student to suspect the presence of symmetry elements.

Having suspected that a symmetry plane may be present in an oblique position, either of two testing procedures may be employed to confirm the presence of the element.

Method 1 is somewhat laborious but is perhaps the clearest way of illustrating the symmetry element, and consists of re-orienting the entire stereogram so that the suspected plane or axis passes through the centre of the diagram. The suspected position of one of the planes is shown in Fig. 3.2(a), intersecting the great circle representing the previously recognized plane *cdef*, at the point *x*. In angular terms, *x* is mid-way between the faces *c* and *d*. Using the net it is now possible to measure the angle between *x* and the centre of the stereogram, in this case 25 degrees. It should now be clear that if the whole stereogram is rotated through 25 degrees about an axis passing through faces *a* and *b*, the plane *axb* will become vertical and pass as a straight line across the diagram. Once in this position it will be easy to see if the plane is a symmetry plane or not.

The rotation is accomplished in the following manner. The diagram is rotated on top of the stereographic net until the faces *a* and *b* coincide with the north and south poles of the net, as shown in Fig. 3.2(b). All the faces are now moved to their new position 25 degrees along the small circle on which they lie. In this case in order to transfer the point *x* to the centre of the stereogram it is necessary to move all points on the upper hemisphere eastwards, i.e., faces *c*, *d*, *g*, and *h* and the point *x*, while faces on the lower hemisphere, *e*, *f*, *i*, and *j*, are moved westwards. In each case the face-pole is moved along the appropriate small circle on the net. Faces *a* and *b* do not move since they coincide with the axis of rotation. Figure 3.2(b) shows the direction and amount of movement for the selected faces *h* and *i*.

When all the faces have been moved to their new positions a new stereogram, shown in Fig. 3.2(c) is obtained. It is clear from this that the plane *aghb* is indeed a plane of symmetry. It also becomes clear at this stage that the third plane of symmetry now lies in the plane of the primitive circle, since each face on the upper hemisphere is now matched by one on the lower hemisphere.

Method 2 saves time compared with the previous method but requires some practice when crystals with a large number of faces are involved. It consists simply of marking the suspected plane on the diagram, in this case the zone *aghbji*, arranging it over the net in the same way as shown in Fig. 3.2(b) and measuring the distance along small circles between the zone and all faces which do not lie on it. The plane

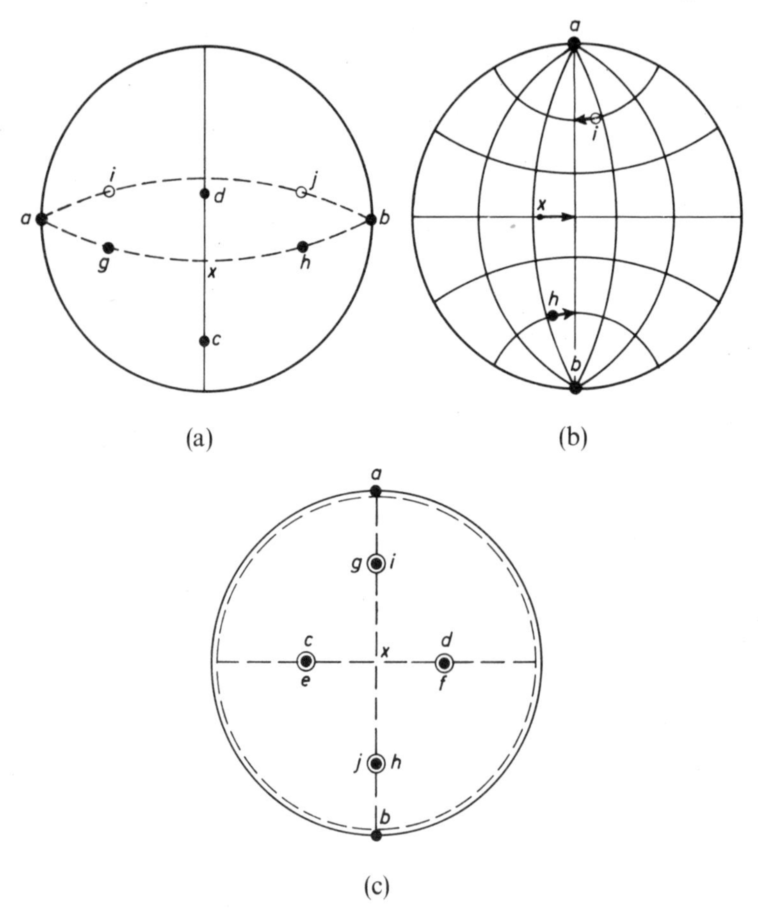

3.2 Partial stereograms of crystal shown in Fig. 3.1. (a) showing the position of one of the suspected symmetry planes (dashed arcs). (b) positioning of the stereogram for rotation about the *ab*-axis, showing also the direction of movement of face-poles on rotation. (c) the stereogram after re-orientation. Broken lines show the positions of the symmetry planes.

concerned is a symmetry plane if every face is matched by another face on the same small circle an equal angular distance the other side of the plane. In our case we find that faces *c* and *d* are each 45 degrees from the plane *aghbji* and are on opposite sides of it. The same applies to faces *f* and *e*. Hence *f* is the mirror image of *e*, and *c* is the mirror image of *d*, and we can be certain that the plane *aghbji* is a symmetry plane.

The procedure in finding symmetry axes is similar in many respects to that outlined above. In Fig. 3.3, for example, a stereogram for an octahedron is given, an example of a crystal belonging to the cubic system and showing at least one obvious four-fold

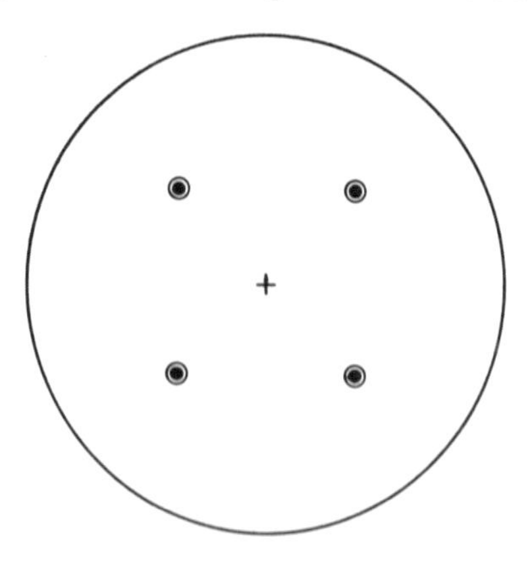

3.3 Stereogram of octahedron with four-fold axis in centre of diagram.

rotation axis, lying in the centre of the diagram. This can easily be demonstrated as a four-fold axis because it is clear that as all the faces of the diagram are rotated about the centre they fall into a position of congruence every 90 degrees, that is to say after 90 degrees of rotation each face-pole falls on the site occupied by another face-pole before the rotation was made. Going back to Fig. 3.2(c) we can now see that the point *x* represents a two-fold axis because in this case it would be necessary to rotate through 180 degrees about *x* before a position of congruence was obtained. Thus, as is the case with symmetry planes, symmetry axes are easy to recognize providing that they coincide with the centre of the diagram.

Suspected axes lying in oblique positions may now be tested by rotating the stereogram so that the axis concerned comes to the centre. In the case of the octahedron, a three-fold axis coincides with each face-pole and the stereogram can be re-oriented to bring one to the centre. The diagram is arranged on the net as in Fig. 3.4(a) and the upper hemisphere face *a* is brought to the centre by rotation of 55 degrees about the axis *mn*. All other faces are moved by the same amount, the movement of the upper face *b* and the lower face *c* being shown on the figure. The new stereogram obtained when all the face-poles have been so moved is shown in Fig. 3.4(b) and makes it clear that face *a* coincides with a three-fold axis.

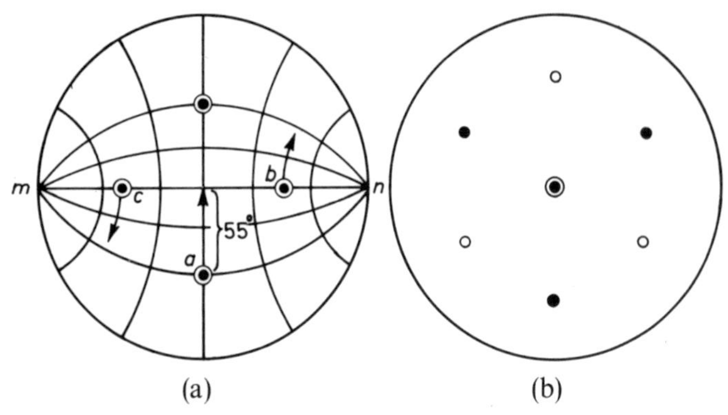

3.4 (a) Rotation of the stereogram of the octahedron to bring a three-fold axis to the centre. (b) The re-oriented stereogram showing the three-fold axis centrally.

The shorter method of testing for the presence of an axis depends on the fact that a set of faces which are symmetrically related to each other by the symmetry operation of the axis, that is, those faces which periodically will occupy each other's position as the stereogram is rotated about the axis concerned, are all angularly equidistant from adjacent members of the set and are all equidistant from the axis. Thus, in the octahedron, measurement of interfacial angles on the stereogram (or indeed on the crystal itself) will show that a face such as A (see Fig. 3.5) lies at 70 degrees from the faces D, B, and *e* while the angles between D and B, B and *e*, and *e* and D are all 110 degrees. Thus as far as the set of faces D, B, *e* is concerned the face A marks the site of a triad axis. It is necessary, however, that *all* the faces of the crystal should show this behaviour. Hence further measurement is necessary to show that each pair chosen from the faces C, *h*, and *f* shows the same interfacial angle and also that all three faces are equidistant from A. The face *g* we are not concerned with because it is opposite A and thus also lies on the axis.

For our present purposes there now remains only the *centre of symmetry* to discuss. A crystal which possesses a centre of symmetry is easily identified in the stereographic projection since every face is matched by another exactly opposite it. In Fig. 3.1, for example, it should be clear that face *c* is opposite face *f*, while *d* is opposite *e*, *h*

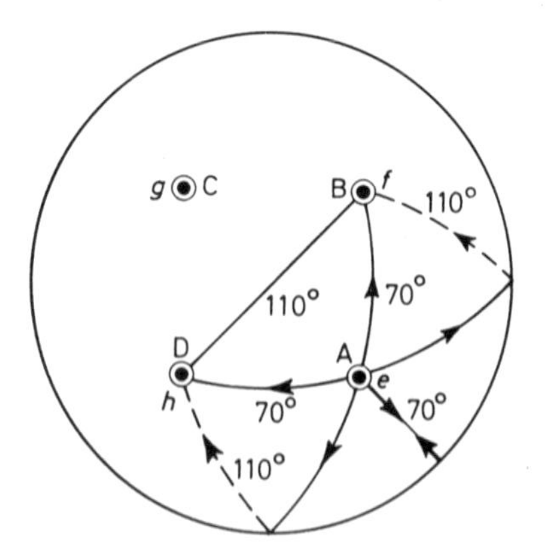

3.5 Interfacial angles in the octahedron. Upper faces are indicated in capitals, lower faces in small letters. Note the technique of measuring the angle between an upper and a lower face, e.g., the 110-degree arc between A and *h*.

is opposite *i*, *g* is opposite *j*, and *a* is opposite *b*. The centre of symmetry is still easily visible in the diagram even when the diagram has been rotated to a new orientation as in Fig. 3.2(c).

Determination of Crystal System

Once the symmetry elements of a crystal have been determined it is normally an easy matter to decide on the crystal system. The character, numbers and types of axes which define the seven systems were given in chapter 1 (p. 5) but it should always be remembered that the axes given are essential axes of symmetry, and a given crystal may have additional axes of other types. For example, the possession of three triad axes immediately marks a crystal as belonging to the cubic system, but many cubic crystals also possess a number of tetrad axes in addition, and all possess additional diad axes. The additional axes, however, do not cause confusion as to the crystal system involved.

Choice of Crystallographic Axes

Once the crystal system has been determined the next step in the full investigation of the crystal is the assigning of crystallographic axes. The following conventions are used:

(a) *Cubic Crystals.* In cases where the three tetrad axes have been identified, the three crystallographic axes a_1, a_2, and a_3 are marked on the stereogram to coincide with them. It does not matter which is which, but it is conventional to orient the stereogram so that a_3 occupies the centre, $+a_1$ lies on the primitive circle at the bottom of the diagram, and $+a_2$ lies on the primitive at the right hand side of the diagram (see Fig. 3.6).

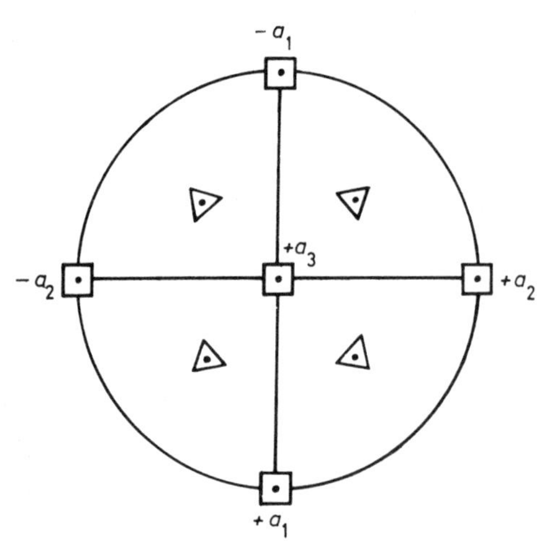

3.6 Conventional orientation of stereograms for crystals of the cubic system. Symbols show positions of four-fold and three-fold axes. The crystallographic axes are a_1, a_2, and a_3.

For cubic crystals which do not possess the three tetrad axes, e.g., the tetrahedron (p. 52) it is necessary to orientate the stereogram so that the triads occupy the same positions as they do in the stereogram of the more symmetrical crystal shown in Fig. 3.6. Note that in the case of the tetrahedron we have an example of crystallographic axes which coincide neither with important symmetry elements nor with face-poles.

(b) *Tetragonal Crystals.* In this system the solitary tetrad axis corresponds with the *c*-crystallographic axis and is arranged to lie at the çentre of the diagram. The remaining axes, a_1 and a_2, lie in the same position as they do for cubic crystals and are mutually perpendicular and normal to the *c*-axis. If a single prism form is present, that is, a set of four faces lying parallel to the *c*-axis and plotting as face-poles on the primitive circle, it is conventional to place the a_1- and a_2-axes diagonally between them as shown in Fig. 3.7. If two prism forms are present, that is to say, eight faces at 45 degrees to each other, the *a*-axis will coincide with alternate face-poles in the primitive. If no prism faces are present there may be several alternative ways of assigning the *a*-axes.

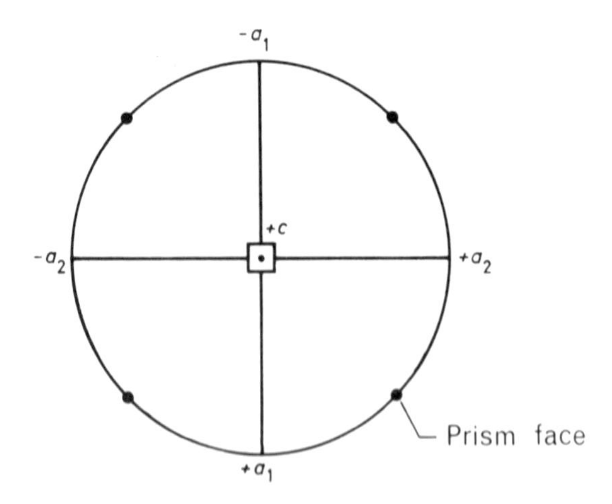

3.7 Conventional orientation of tetragonal stereograms.

(c) *Orthorhombic Crystals*. Orthorhombic crystals possess three diad axes which coincide with the crystallographic axes *a*, *b*, and *c*. For present purposes it does not matter which crystallographic axis is assigned to which diad but a further discussion of this point is given on p. 58. Figure 3.8 shows a typical orthorhombic stereogram with the crystallographic and symmetry axes marked.

(d) *Monoclinic Crystals*. In this system a single diad is present and coincides with the *b*-crystallographic axis. The *a* and *c* axes lie in a plane normal to the *b*-axis, which is in many common crystals, e.g., augite, hornblende, and gypsum, also a plane of symmetry. The symmetry plane is arranged to run north–south across the diagram, while +*b* lies on the primitive on the right-hand side (see Fig. 3.9). The positioning of *a* and *c* within the plane normal to *b* is often simply a matter of convenience and consists of choosing two directions such that as many faces as possible are parallel to the axes chosen. If a prominent form of prismatic appearance is present this is taken to be parallel to the *c*-axis. The conventional orientation is then to place *c* in the centre of the diagram while the face-poles of the prism form

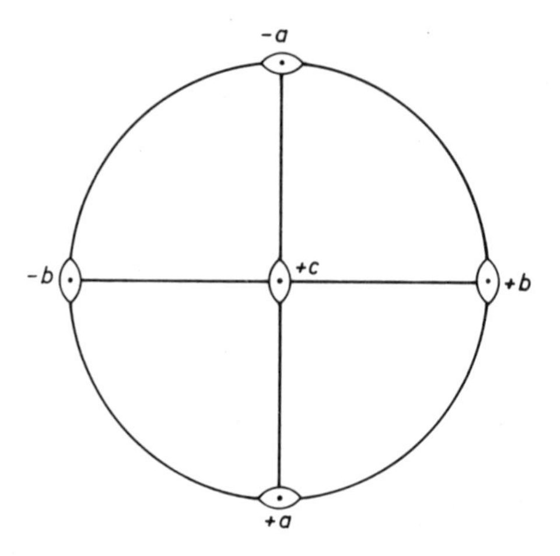

3.8 Conventional orientation of orthorhombic stereograms.

plot round the primitive circle. The *a*-axis is then taken provisionally to be parallel to the most prominent face or edge remaining in the zone normal to *b* (a further discussion of the choice of axes is given

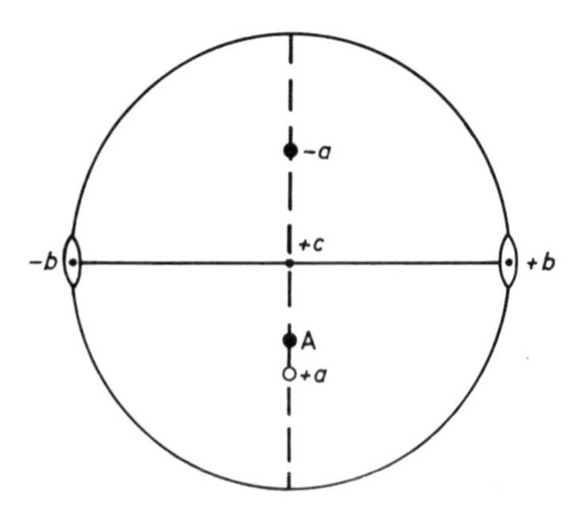

3.9 Conventional orientation of monoclinic stereograms. A is the face-pole of a prominent face or edge chosen to be parallel to the *a*-crystallographic axis. The angle β lies between $+c$ and $+a$ (the latter on the lower hemisphere), the angle A $\wedge$ $+a$ is 90 degrees.

on p. 39. Such a face will in general not be at right angles to the c-axis and its face-pole will not therefore coincide with the centre of the stereogram. The convention is to orientate the stereogram so that the face parallel to the a-axis plots south of the centre of the diagram (see Fig. 3.9) with the consequence that the positive end of the a-axis plots on the *lower* hemisphere, in the southern half of the stereogram, while the negative end plots on the upper hemisphere in the northern part of the stereogram.

(e) *Trigonal Crystals*. Crystals of the trigonal system possess a single triad axis as their essential element of symmetry. This coincides with the c-crystallographic axis, and is plotted centrally in the

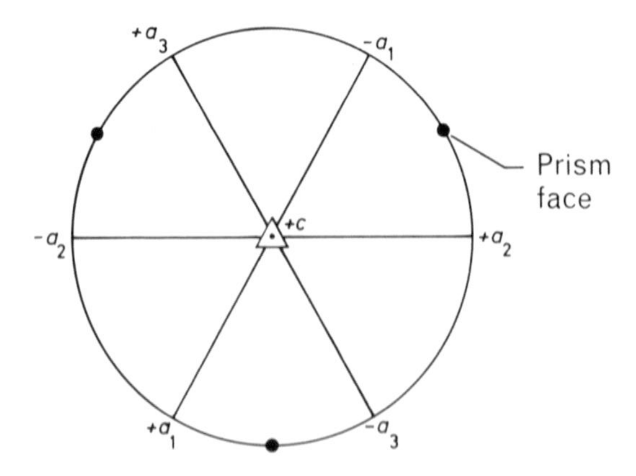

3.10 Conventional orientation of trigonal stereograms.

stereogram. The three a-axes lie normal to c, and plot in the primitive circle each one 120 degrees from the next. Note that this results in alternate positive and negative ends of the axes appearing round the primitive circle. If a single prism form is present, i.e., a form consisting of three faces each 120 degrees apart, plotting in the primitive circle, the a-axes are arranged to be parallel to the faces (see Fig. 3.10) so that the face-poles of the prism faces fall half-way between the crystallographic axes.

(f) *Hexagonal System*. The single hexad of this system corresponds with the c-crystallographic axis and is plotted in the centre of the stereogram. The a-axes are arranged as in the trigonal system (see Fig. 3.11). As in the tetragonal and trigonal systems, when a single prism form is present, in this case six faces, the face-poles plot midway between the ends of the axes.

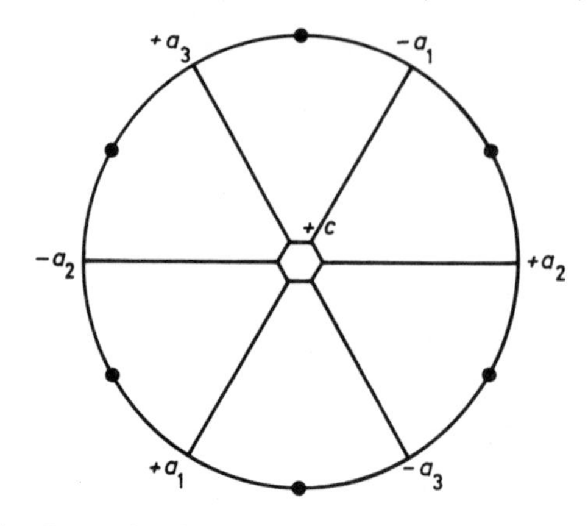

3.11 **Conventional orientation of hexagonal stereograms.**

(g) *Triclinic Crystals.* Triclinic crystals have no axes of symmetry and three crystallographic axes *a*, *b*, and *c*, are chosen so that as many faces as possible are parallel to two axes. This results in a simple set of Miller indices being obtained. The conventional orientation is to position *c* in the centre of the diagram (see Fig. 3.12).

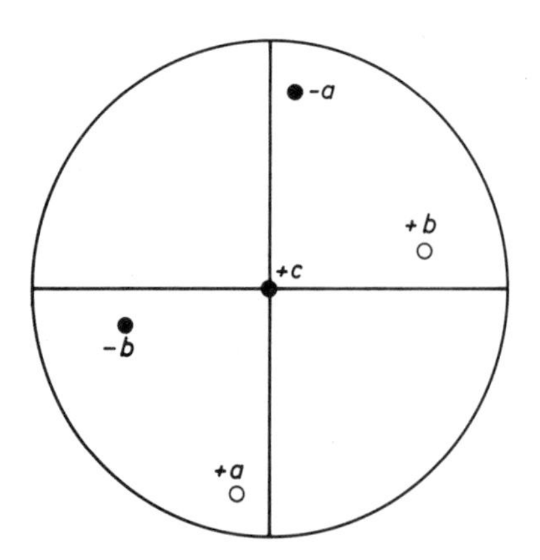

3.12 **Conventional orientation of triclinic stereograms.**

Determination of Angles between Axes

In the monoclinic and triclinic systems the axes are at variable angles depending on the crystal concerned. The determination of these angles may aid in identifying the substance and is easily carried out using the stereographic net. Two axes are arranged to lie on a single great circle and the angular distance between them is measured in the usual way. The result for monoclinic crystals is usually expressed as the obtuse angle β, that is the angle between $+a$ and $+c$. The angles between a and b, and c and b are of course right angles. In triclinic crystals it is necessary to measure all the angles between each pair of axes and these are expressed as α (between b and c), β (between a and c), and γ (between a and b).

It must always be remembered, however, that for every mineral species there is an accepted orientation of the crystallographic axes and the choice made by an individual working with an unidentified mineral may not be the same as is usually accepted for the mineral concerned. Hence, determinations of angles between axes may yield results which differ from those given in standard texts, with the result that identification is made difficult. The problem is, however, much less acute than it seems, for it is easy to try the few alternative arrangements of axes which the stereogram will allow and see if better results can be obtained.

Choice of Parametral Plane

As an essential preliminary to the indexing of faces and the determination of axial ratios, it is necessary to choose a face as a parametral plane, that is a reference face to which others may be related (see p. 12). The parametral plane cuts the positive ends of all three axes in the cubic, tetragonal, orthorhombic, monoclinic, and triclinic systems and hence in general lies in the lower right-hand quadrant of the conventionally oriented stereogram. In the trigonal and hexagonal systems the parametral plane makes equal intercepts on the positive end of the a_1-axis and the negative end of the a_3-axis, while being parallel to a_2.

The parametral plane in any system, however, must always be a *unique* face within the solid angle made by the nearest crystallographic axes, e.g., in the cubic, tetragonal, and orthorhombic systems it must be the only face of its form lying in the lower right-hand quadrant of the stereogram. This means in practice that the parametral plane must *lie on* any symmetry plane or axis which runs through the quadrant. If it did not, one can easily see that there would be more than one face of the same kind within the quadrant, and hence the condition of uniqueness would not be fulfilled (see Fig. 3.13).

Within the limitation expressed, however, there is still some choice left in the positioning of the parametral plane in all systems except the cubic. In many crystals belonging to the latter system several symmetry planes run through the quadrant and intersect in a triad axis. Only the face which *coincides* with the triad axis is acceptable as the parametral plane (the octahedron face).

In other systems, such as the trigonal, hexagonal, and tetragonal, a single symmetry plane may run through the solid angle concerned. In these cases any face whose face-pole plots on the symmetry plane is a possible parametral plane (see Fig. 3.13).

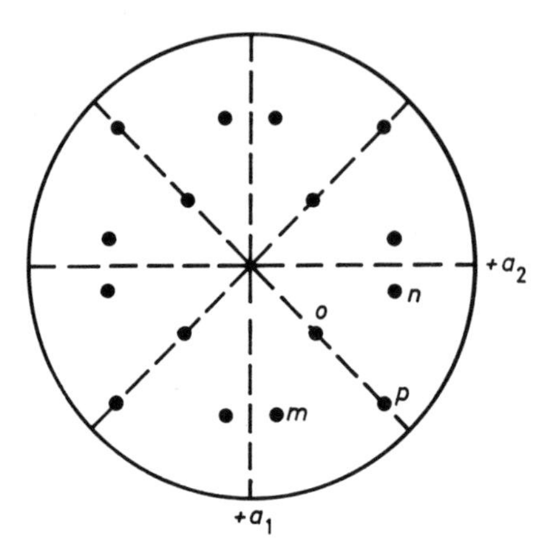

3.13 Choice of parametral plane illustrated in a tetragonal stereogram (symmetry planes shown by broken lines). Either of the faces *p* and *o* could be used since both are unique within the quadrant. Neither *m* nor *n* could be used because they are related by a symmetry plane within the quadrant and are hence not unique.

In monoclinic, triclinic, and orthorhombic crystals no symmetry elements are present within the solid angle between the positive ends of the axes, and hence any face within the solid angle may theoretically be chosen.

This raises two points. Firstly, as was the case with the angles between axes mentioned previously, every mineral species has a particular face which is generally accepted as being the parametral plane. Thus, although theoretically correct, the face chosen by an individual working with an unknown mineral may not coincide with the accepted parametral plane. This will result in a non-standard set of Miller indices being assigned to the faces, and in the determination of 'incorrect' axial ratios. Again, the problem is not serious because the number of alternative choices is usually small and it is a rapid procedure to calculate alternative axial ratios and alternative sets of Miller indices.

The second point concerns the procedure when the crystal possesses no faces which are possible choices as parametral plane (for example any cubic crystal which does not carry either octahedral or tetrahedral faces). In this case it is necessary to calculate the position of a suitable face-pole by the methods outlined on p. 44, and choose this hypothetical face as the parametral plane. The cube itself provides a very simple example of this technique, since despite the fact that no octahedral face is present it is possible to calculate exactly the positions of the triad axes, which, as has been shown, must coincide with the octahedral face poles, one of which must be taken as the parametral plane.

Indexing of Faces

Now that the parametral plane and the crystallographic axes have been chosen, the faces of the crystal may be given Miller indices. We shall discuss first the crystal systems in which there are three crystallographic axes.

The first step is to write on the index (111) of the parametral plane and then proceed to carry out all the symmetry operations which the symmetry elements of the crystal require. In other words, using the net, it is necessary to find out where the face (111) is repeated by reflection in any symmetry planes present, and by rotations of the appropriate number of degrees about any symmetry axes present. When this step is completed, depending on the crystal system, up to seven other faces related by symmetry operations to (111) will have been found. These constitute all the faces of the *form* {111} and they will all have indices of this type. The symbol {111} is used to indicate collectively all the faces which have the same relations to the crystal lattice as the face (111). The entire set of such faces is referred to as a form. The number of faces in a form depends on the symmetry elements present. Inspection of the stereogram will now show how to distribute the minus signs among the indices. In Fig. 3.14 an indexed stereogram of the upper half of an octahedron is given and shows the faces (111), ($\bar{1}11$), ($1\bar{1}1$), and ($\bar{1}\bar{1}1$).

The second step is to index the faces which are parallel to two crystallographic axes, remembering that, because of the reciprocal relationship in the derivation of indices, the symbol 0 signifies an infinite intercept on an axis, i.e., parallelism with that axis. Thus in Fig. 3.15, a stereogram of a cube, the face x is parallel to the a_2- and

a_3-axes and cuts the positive end of the a_1-axis. It thus receives the indices (100), while the face opposite it, *y*, receives the indices ($\bar{1}00$). The indexing of the other faces should be clear from the figure.

Step three consists of drawing zones (great circles) where necessary to join up the remaining unindexed faces to those which have already

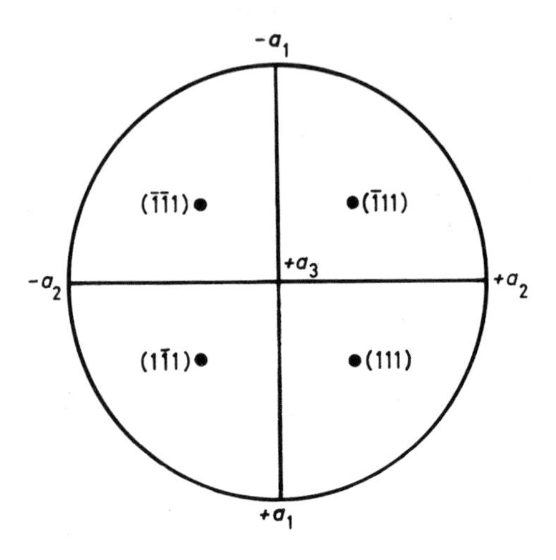

3.14 Miller indices for the upper faces of the octahedron.

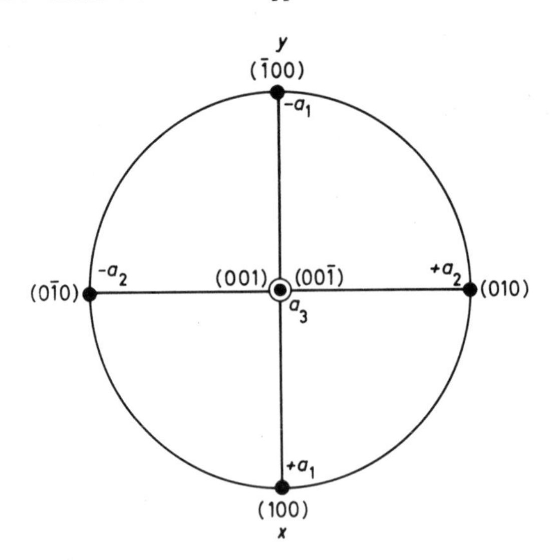

3.15 Miller indices for the faces of the cube.

been indexed. It is essential to find two zones which pass through each unknown face (see Fig. 3.16). It is now possible to use one of the interesting geometrical properties of the stereographic projection to find the indices of the unknown face. In any zone, adding together the Miller indices of two faces, providing that they do not lie opposite to each other, gives the indices of a third face which lies on the zone between them. This theorem, it is most important to note, does not state *precisely* where the third face lies, but only that it lies between

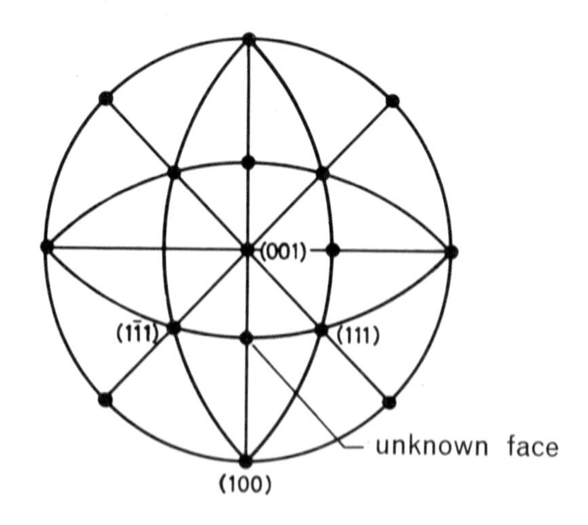

3.16 Indexing an unknown face. For explanation see text.

the other two. However, in the example given in Fig. 3.16 the unknown face lies on two zones:

(a) between (100) and (001)
(b) between (111) and $(1\bar{1}1)$

Adding the indices in case (a) we obtain (101), while in case (b) we obtain (202), which is of course the same as (101). Thus the face (101) lies on *both* zones and must therefore lie at their intersection. Hence the unknown face is (101).

It cannot be stressed too strongly that *two* zones must give the same answer before the identity of the unknown is determined. If the two zones chosen give different answers it is always worth inspecting the stereogram to see if a different pair of zones could be employed. Frequently it is possible to index all the faces of the crystal by working out the indices of zone intersections in all the cases where the addition of indices gives a concordant result for both zones, and then, using the indices thus determined, to draw further zones through these points. The new zones provide new possibilities for intersections, and the process may be continued until all the unknown points have been indexed. It may be necessary during the calculation, to index points which are not represented as faces on the crystal under study. All these are of course *possible* faces.

The method of adding indices in many cases gives the indices of an unknown face quickly and easily. There are occasions, however, when it will be found more convenient to use a more elaborate calculation, which is somewhat lengthy but has the advantage of giving a solution in all cases. The calculation consists of two steps, firstly calculating the zone symbol of zones which are defined by two known faces. As in the previous method it is required to draw two zones through the unknown face, each zone being already defined by two known faces which must not be opposite each other. The next step is to calculate the symbol of the *zone axis* (i.e., the line to which all faces in the zone are parallel) of each zone. This calculation is carried out as follows: the indices of one of the known faces in the zone are written down twice on one line, while the indices of another known face are written below. Thus for the case considered in Fig. 3.16 the first zone (between (100) and (001)) is written as follows:

$$\begin{array}{c|cccc|c} 1 & 0 & 0 & 1 & 0 & 0 \\ & & \times & \times & \times & \\ 0 & 0 & 1 & 0 & 0 & 1 \end{array}$$

The first and last columns are then excluded and the remaining four columns are cross-multiplied in the following way: The upper figure in the first remaining column is multiplied by the lower figure in the next column. From this product is subtracted the result of the multiplication of the upper figure in the second column with the lower figure in the first. This gives the first digit in the Miller indices of the zone axis. In this case we have:

$$0 \times 1 - 0 \times 0 = 0$$

The second digit is obtained in the same way, but by cross-multiplying the second and third columns, while the third digit is obtained from the third and fourth columns:

$$\begin{array}{lll} \text{2nd digit} & \ldots & 0 \times 0 - 1 \times 1 = \bar{1} \\ \text{3rd digit} & \ldots & 1 \times 0 - 0 \times 0 = 0 \end{array}$$

Hence the indices for the zone axis, usually known as the *zone symbol*, are $[0\bar{1}0]$ which is the same direction as $[010]$, the square bracket being used to distinguish the indices from those of the face-pole, or form, e.g., (010) and $\{010\}$.

It is now a simple matter to repeat this procedure with the second zone, that which includes (111) and $(1\bar{1}1)$. Care is needed to keep the signs correct when faces with negative indices are used. In this case we obtain:

$$\begin{array}{c|cccc|c} 1 & 1 & 1 & 1 & 1 & 1 \\ & \times & \times & \times & & \\ 1 & \bar{1} & 1 & 1 & \bar{1} & 1 \end{array}$$

1st digit ... $1 \times 1 - 1 \times \bar{1} = 2$
2nd digit ... $1 \times 1 - 1 \times 1 = 0$
3rd digit ... $1 \times \bar{1} - 1 \times 1 = \bar{2}$

The zone symbol is therefore $[20\bar{2}]$, i.e., $[10\bar{1}]$.

The final step in finding the indices of the unknown face lying at the intersection of the two zones consists of following the same procedure again, this time using the two sets of zone symbols:

$$\begin{array}{c|cccc|c} 0 & 1 & 0 & 0 & 1 & 0 \\ & \times & \times & \times & & \\ 1 & 0 & \bar{1} & 1 & 0 & \bar{1} \end{array}$$

giving:

1st digit ... $1 \times \bar{1} - 0 \times 0 = \bar{1}$
2nd digit ... $0 \times 1 - 0 \times \bar{1} = 0$
3rd digit ... $0 \times 0 - 1 \times 1 = \bar{1}$

Thus the indices of the unknown face are $(\bar{1}0\bar{1})$, which is the same plane as (101), the answer derived previously.

This type of calculation may seem a very roundabout way of deriving the same answer as had already been obtained so easily by the addition method, but it must be remembered that the cross-multiplication method will yield a correct solution in all cases (providing the arithmetic has been done correctly!) while the simpler method will only give the answer in favourable cases.

With the information now available the student should be able to index every face on the crystal (in cubic, tetragonal, orthorhombic, monoclinic, and triclinic examples), no matter how complex its indices.

Indexing Hexagonal and Trigonal Crystals

The four-digit symbols of the Miller–Bravais system were mentioned on p. 15. Indexing the faces of hexagonal and trigonal crystals is in principle very similar to indexing crystals of the other systems but note that the parametral plane is $(10\bar{1}1)$. The calculation of indices of unknown faces, expressed in general as $(xyuz)$, proceeds with the omission of the symbol u (representing the intercept on the a_3-axis) from the indices of the known faces used in the calculation. At the end of the calculation a new value for u may be re-inserted because of the relationship $x + y + u = 0$. The following example should make this point clear:

Given two zones, the first of which includes $(10\bar{1}2)$ and $(10\bar{1}1)$, the second $(1\bar{1}00)$ and $(0\bar{1}10)$, what face lies at their intersection?

The indices of the first zone are changed to (10*2) and (10*1), where the asterisk represents the omitted u symbol, and, similarly, the faces of the second zone become $(1\bar{1}{*}0)$ and $(0\bar{1}{*}0)$. The calculation of the zone axis for each zone then proceeds in the usual way using the three digits available and gives the results [01*0] and [00*1]. These zone symbols are complete in themselves and it is *not* correct to calculate a new value of u and re-insert it at this stage. The indices of the unknown face are then derived by the cross-multiplication of the two zone symbols, which gives the answer [10*0]. The value of u can now be calculated as -1 and re-inserted to give $(10\bar{1}0)$ as the full answer.

Determination of Axial Ratios

Once a parametral plane has been chosen it is possible to work out the axial ratios of the crystal. These ratios may be helpful in identification of the substance but as was pointed out on p. 40, because of the element of choice available for the parametral plane, the axial ratios determined by the individual may differ from those usually accepted for the substance. If no correspondence is found between the ratios determined and published ratios another parametral plane should be chosen and new ratios calculated. In cases where the identity of the substance is already known, the calculation of axial ratios and the comparison with standard data will help the investigator to find the standard orientation and parametral plane for his specimen.

A general formula for the calculation of axial ratios is derived as follows. Figure 3.17 shows three crystallographic axes a, b, and c, intersecting at the point O. The angles between the axes are not important. The plane ABC represents the parametral plane, the intercepts on the three axes being AO, BO, and CO. The line OM is the normal to the parametral plane, that is the face pole of (111),

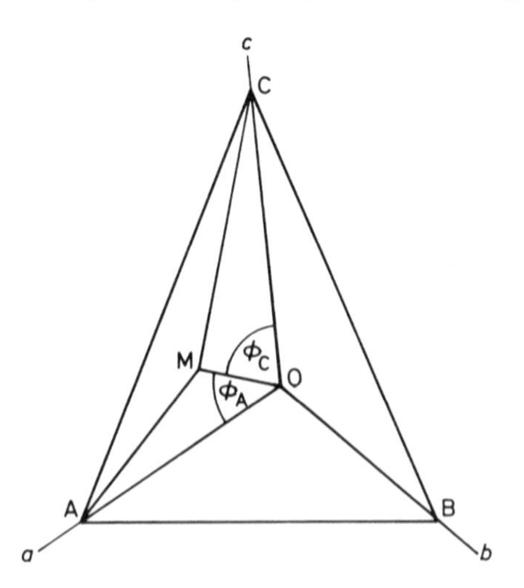

3.17 Calculation of axial ratios from the stereogram. For explanation see text.

and cuts the plane ABC at M. The angles between each axis and the line OM are represented by ϕ_A, ϕ_B, and ϕ_C. For clarity the angle ϕ_B, the angle BOM, is not shown on the diagram. Then:

$$\frac{OM}{OC} = \cos\phi_C \quad \text{and} \quad \frac{OM}{OA} = \cos\phi_A \quad \text{and} \quad \frac{OM}{OB} = \cos\phi_B$$

Hence the axial ratios expressed as

$$\frac{a}{b} : \frac{b}{b} : \frac{c}{b}$$

are given as:

$$\frac{OA}{OB} : \frac{OB}{OB} : \frac{OC}{OB} = \frac{\cos\phi_B}{\cos\phi_A} : 1 : \frac{\cos\phi_B}{\cos\phi_C}$$

which is more conveniently expressed as:

$$\frac{a}{b} : \frac{b}{b} : \frac{c}{b} = \sec\phi_A : \sec\phi_B : \sec\phi_C$$

In any of the crystal systems employing three crystallographic axes the angles ϕ_A, ϕ_B, and ϕ_C, are easily measured on the stereogram. Figure 3.18 shows where they lie in an orthorhombic stereogram as an example.

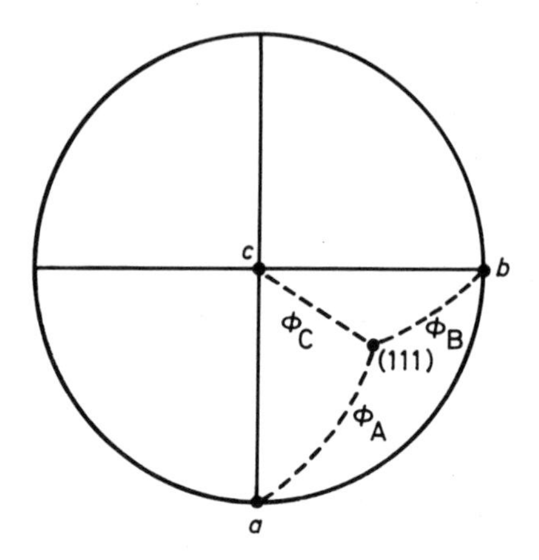

3.18 Calculation of axial ratios. For explanation see text.

In the various crystal systems the following general relationships are found between the angles ϕ_A, ϕ_B, and ϕ_C:
Cubic $\phi_A = \phi_B = \phi_C$ and the axes are therefore all equal.
Tetragonal $\phi_A = \phi_B \neq \phi_C$ so that the a_1- and a_2-axes are equal.
Orthorhombic, *monoclinic*, and *triclinic* $\phi_A \neq \phi_B \neq \phi_C$ so that all the axes are unequal.

In the trigonal and hexagonal systems the three *a*-axes are all equal and determination of the axial ratios only involves finding the ratio *c*/*a*. The calculation is the same in principle as that outlined above, but in this case we require to know the angle between the face-pole of the parametral plane and the *c*- and a_1-axes respectively. Figure 3.19 shows where these lie on the stereogram, and by a modification of the previous calculation the student should easily be able to demonstrate that:

$$\frac{c}{a} = \frac{\cos \phi_A}{\cos \phi_C}$$

The method outlined above for the determination of axial ratios provides a relatively simple means of performing this calculation. It should be noted, however, that the method depends on the determination of values of ϕ, by measurement of the stereogram with the stereographic net. The accuracy with which the axial ratios are determined thus depends not only on the accuracy of the original measurements of interfacial angles but also on the accuracy with which it is possible to construct and measure the stereogram.

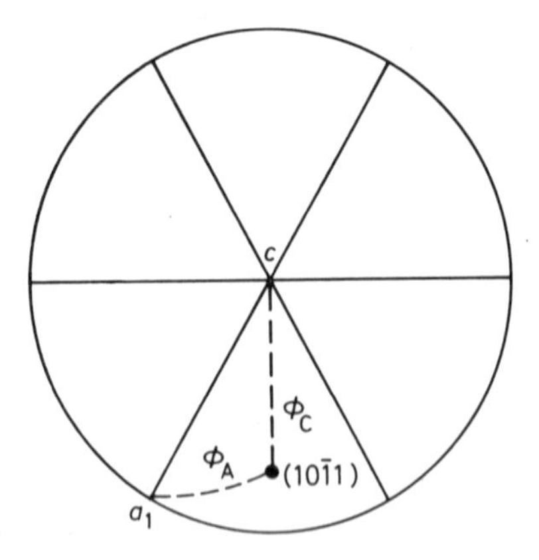

3.19 Calculation of axial ratios in trigonal and hexagonal crystals. For explanation see text.

Advanced texts on crystallography give methods of determining axial ratios which depend only on the measured interfacial angles and can thus be of great accuracy if accurate measurements have previously been obtained (e.g., by methods of optical goniometry).

Fortunately, however, it is rarely necessary nowadays to obtain accurate crystallographic measurements in order actually to identify a specimen. Unknown minerals are usually identified by X-ray diffraction methods and crystallographic measurements of the type dealt with in this book have their main application in identifying particular planes, within minerals whose identity is already known.

4. Systematic crystallography

Introduction

In the preceding chapters the general principles of morphological crystallography and the stereographic projection have been outlined. In this chapter a brief but systematic account of the seven crystal systems and the phenomenon of twinning will be given, with illustrations drawn from the common minerals.

Nomenclature of Faces

In the sections which follow, the nomenclature employed for the faces of crystals is discussed system by system. It is possible, however, to make a few generalizations on the subject which may be helpful at this stage.

Consider first the orthorhombic, monoclinic, and triclinic systems which share the feature of having three different crystallographic axes. Faces which lie parallel to any *two* of the crystallographic axes are generally referred to as *pinacoids*, e.g., (100), (010), and (001). The last face referred to, (001), is more specificially known as the *basal pinacoid.* Faces which lie parallel to the *c*-axis but are not parallel to either of the other axes are known as *prisms*, and while there can only be three different pinacoidal forms there may be any number of prisms such as {110}, {120}, {130}, etc.

Analogous to prisms are the faces which are parallel either to the *a*-axis or to the *b*-axis, e.g., {011} and {101}. Those belong to the general category of *domes*.

Lastly, faces such as (111) which cut all three axes are known as *pyramids*, because when all the faces of the form are developed, a pyramidal shape may be created.

Tetragonal crystals are of higher symmetry than those of the three systems discussed, and not only forms of the {111} type but also forms of the {101} type, produce a pyramidal shape. Hence all the faces which cut the *c*-axis (except the basal pinacoid) come under the general heading of pyramids, and the term dome is not used. By an extension of this, *all* the faces parallel to the *c*-axis are termed prisms, and the term pinacoid is reserved for the basal face, (001), only.

In the hexagonal and trigonal systems the term prism is similarly used for all the faces parallel to *c*, and the only pinacoid is the basal plane (0001). A number of different terms is used for faces cutting *c* in these systems, and these will be considered at a later stage.

Cubic System

The essential symmetry of the cubic system consists of four triad axes, but holosymmetrical crystals of this system possess three tetrad axes, six diad axes, nine symmetry planes, and a centre of symmetry, in addition. The arrangement of these symmetry elements and their relations to the crystallographic axes have been discussed on p. 34.

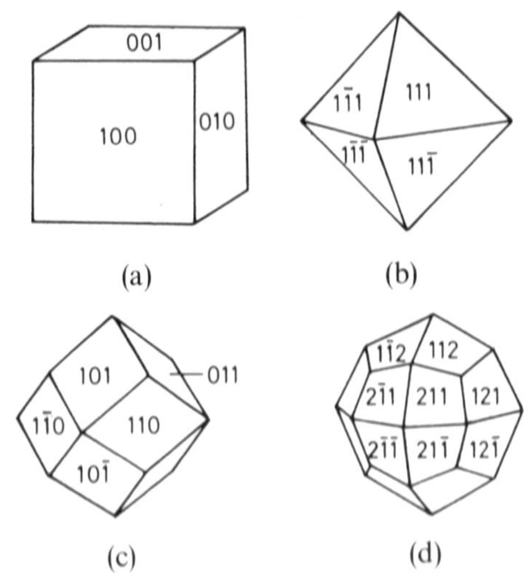

4.1 Common forms of the cubic system. (a) cube; (b) octahedron; (c) rhombdodecahedron; (d) trapezohedron.

The following forms are commonly encountered in holosymmetrical crystals and are illustrated in Fig. 4.1:

(a) Cube {100}, six faces

(b) Octahedron {111}, eight faces
(c) Rhombdodecahedron {110}, twelve faces
(d) Trapezohedron {211}, twenty-four faces

All these forms are illustrated in the stereogram in Fig. 4.2.

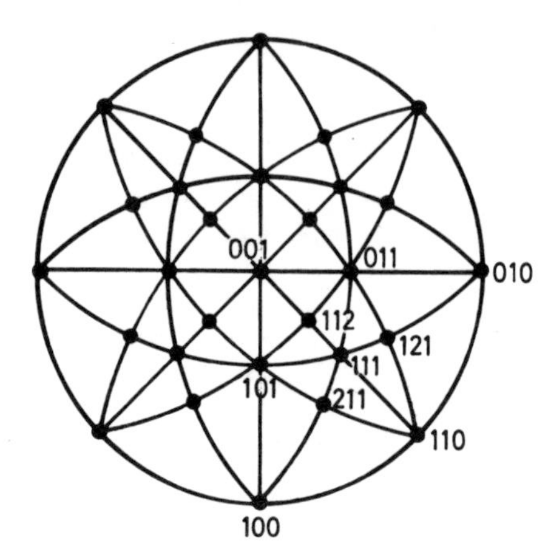

4.2 Stereogram of the forms illustrated in Fig. 4.1 (upper faces only).

Amongst the lower order of symmetry classes in the cubic system the most important forms are the tetrahedron {111} and the pyritohedron {210}.

The *tetrahedron*, illustrated in Fig. 4.3, has faces parallel to those of the octahedron but only half of them are developed. Thus, while retaining the four triads characteristic of the system, the tetrahedron has no tetrad axes, three diads only (which correspond with the crystallographic axes), and only six planes of symmetry.

Tetrahedra may be of two sorts, the one illustrated being the *positive* tetrahedron {111}. This may be accompanied in the same

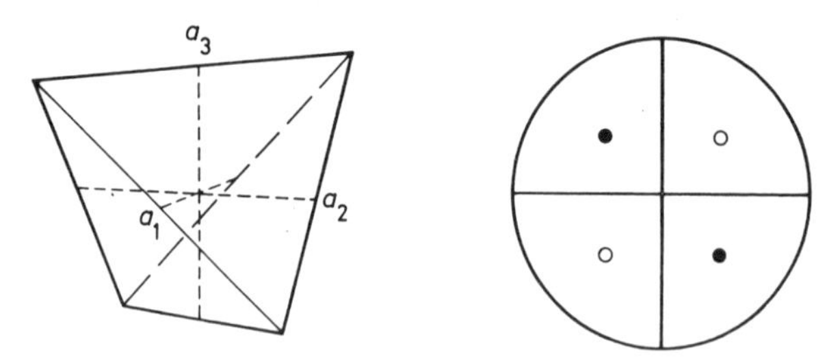

4.3 The tetrahedron.

crystal by the *negative* tetrahedron $\{1\bar{1}1\}$. Equal development of the two forms happening by chance in one crystal will of course lead to the appearance of an octahedron, but the true symmetry is lower than this.

The *pyritohedron* is illustrated in Fig. 4.4 and consists of twelve five-sided faces. The symmetry of this form consists of four triad

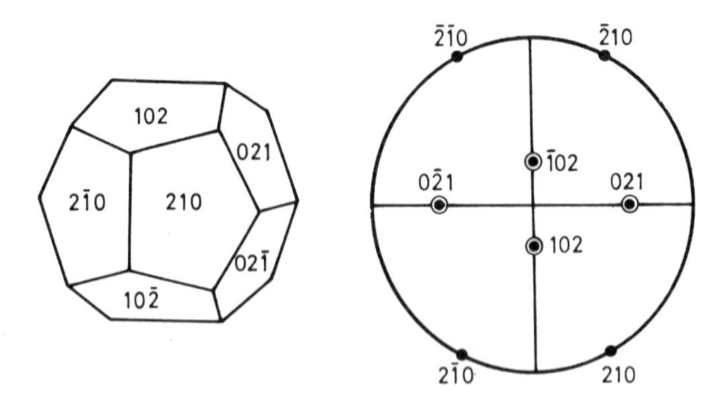

4.4 The pyritohedron.

axes, three diads, three planes, and a centre of symmetry. This form is characteristic of the mineral *pyrite* (p. 104), which also forms crystals of cube-like appearance. The latter, however, consist of two different three-faced forms, as can be seen by the frequent striations on the faces (see Fig. 4.5). It should be clear from the figure that the striated cube does not possess the full symmetry of the cube, but only the same reduced number of symmetry elements as are present in the pyritohedron.

Other Cubic Minerals. Good crystals are obtained of *galena* (p. 103) which is found as cubic crystals often combined with the octahedron.

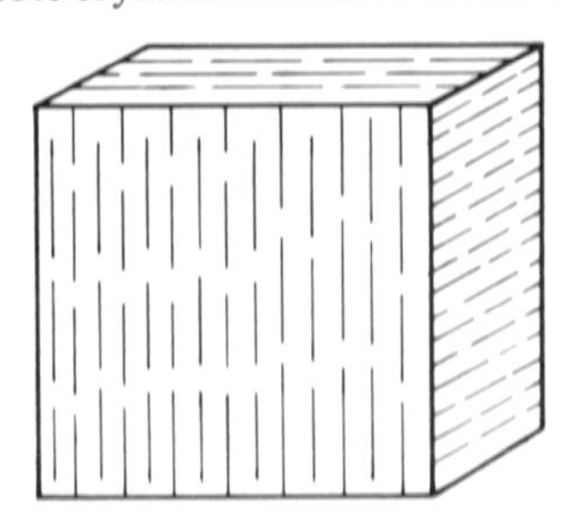

4.5 A striated cube of pyrites.

The excellent cleavage in three directions at right angles (parallel to the faces of the cube) is also characteristic of the mineral.

Fluorite (p. 106) also crystallizes as cubes but has an octahedral cleavage (four directions parallel to the faces of the octahedron) which truncates the corners of the cube.

Garnet (p. 135) is very frequently found in excellent crystals of rhombdodecahedral form, {110}, or trapezohedral form, {211}. The crystal illustrated in Fig. 7.25 shows a combination of both forms.

Leucite (p. 126) and *analcite* (p. 128) also crystallize characteristically as trapezohedra.

Tetragonal System

The essential symmetry of this system consists of a single tetrad axis coinciding with the *c*-crystallographic axis. In the holosymmetrical class there are in addition four horizontal diad axes, four vertical planes of symmetry (i.e., parallel to the *c*-axis), one horizontal plane, and a centre of symmetry. The arrangement of the a_1- and a_2-crystallographic axes has been discussed on p. 34.

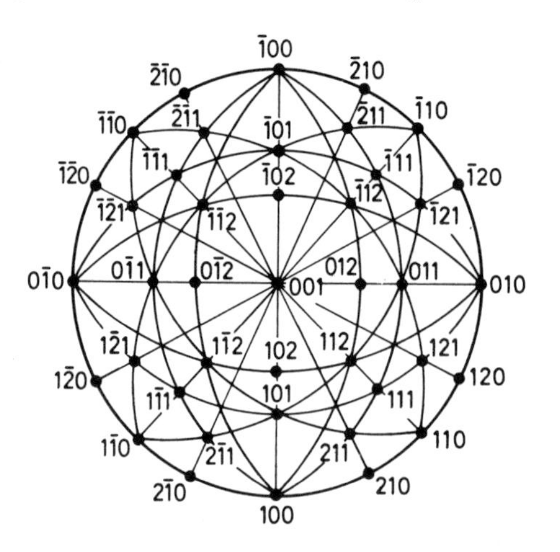

4.6 Stereogram showing some common forms in the tetragonal system (upper faces only).

Common forms in the holosymmetrical class are illustrated in a stereogram in Fig. 4.6 and include:

(a) The prism of the first order {110} (four faces).
(b) The prism of the second order {100} (four faces).

(c) A variety of *ditetragonal* prisms such as {210}. The ditetragonal prisms have faces which do not project onto symmetry planes in the stereogram and hence become eight-faced forms.
(d) The basal pinacoid {001} (two faces).
(e) First-order pyramids such as {111} and {112} (eight faces).
(f) Second-order pyramids such as {101} and {102} (eight faces).
(g) A variety of ditetragonal pyramids such as {211} (sixteen faces).

It should be mentioned that tetragonal crystals are of two types—those which have the axial ratio $c/a > 1$ and those which have $c/a < 1$. It is easy to distinguish these types on the stereogram by measuring the angle between (001) and (101). In crystals where c is longer than a this angle is more than 45 degrees while in crystals in which c is shorter than a it is less than 45 degrees.
Tetragonal Minerals. Tetragonal minerals are not particularly common, and in fact none of the important rock-forming minerals belongs to this system, though amongst other minerals *zircon* (p. 146), *idocrase* (p. 137), and *chalcopyrite* (p. 102) form suitable examples.

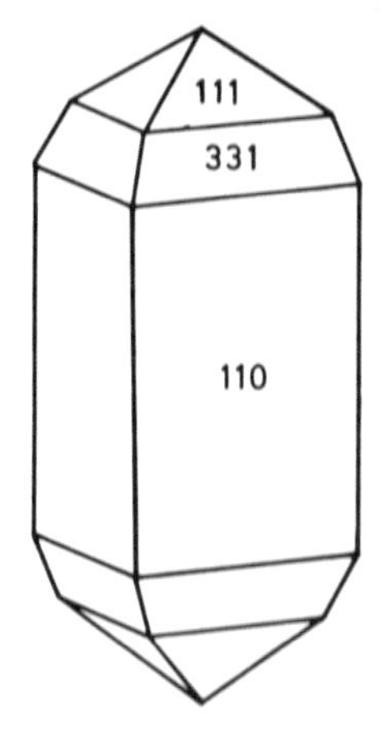

4.7 Zircon, showing first order prism {110}, and two pyramids {111} and {331}.

Zircon crystals are holosymmetrical and have the axial ratio $c/a = 0{\cdot}64$. The crystals commonly consist of a combination of the first-order prism and the first-order pyramid {111}, sometimes accompanied by the steeper first-order pyramid {331} (see Fig. 4.7).
Idocrase has $c/a = 0{\cdot}54$ and is found commonly as a combination

of the first- and second-order prisms with the {111} pyramid and the basal pinacoid (see Fig. 4.8).

Chalcopyrite belongs to a class having a lower order of symmetry and characteristically crystallizes in a form known as a *sphenoid.*

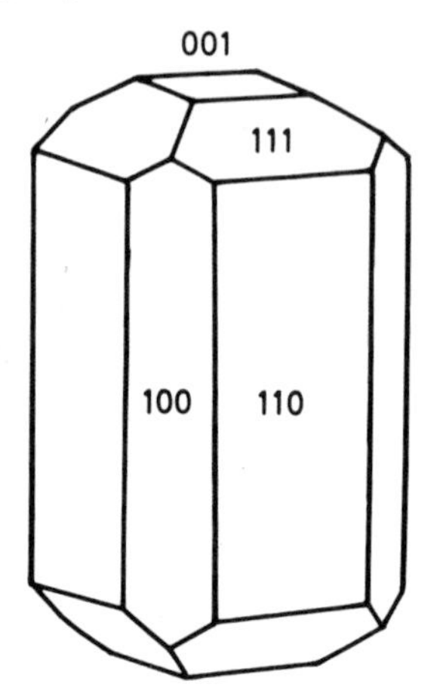

4.8 Idocrase, showing first order prism {110}, second order prism {100}, pyramid {111}, and basal pinacoid {001}.

The latter is akin to the tetrahedral form of the cubic system and is a four-faced form which one may imagine to be produced by the development of alternate faces of the tetragonal first-order pyramid. Thus, as with the tetrahedron, sphenoids may be of two types, the negative sphenoid $\{1\bar{1}1\}$ and the positive sphenoid {111}. The crystal of chalcopyrite illustrated in Fig. 4.9 is a combination of the two

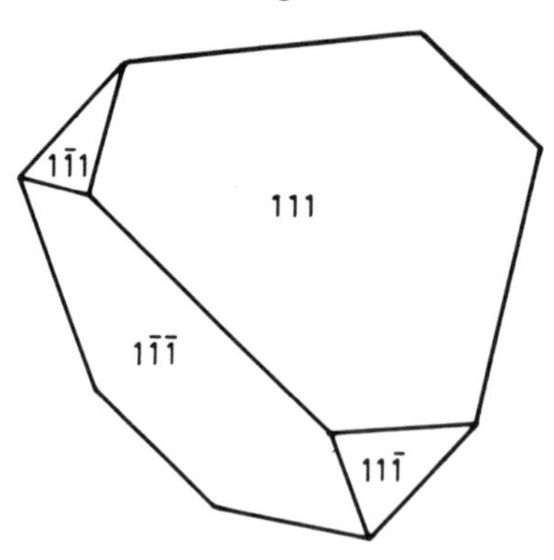

4.9 Chalcopyrite, showing a combination of the positive sphenoid {111} with the negative sphenoid $\{1\bar{1}1\}$.

sphenoids and, since the ratio $c/a = 0{\cdot}99$, such crystals are very similar to slightly irregular octahedra. As is the case with the striated cubes of pyrite previously mentioned, striations on the faces may reveal the low order of symmetry.

Chalcopyrite also serves to exemplify a type of symmetry element which it has so far been convenient to neglect, that is to say the *inversion axis*. Such axes are only important in crystals of low symmetry classes. The symmetry operation concerned involves the inversion of the crystal in order to obtain a position of congruence after every rotation operation. A sphenoid is shown in the stereogram in Fig. 4.10 where a four-fold inversion axis lies in the centre of

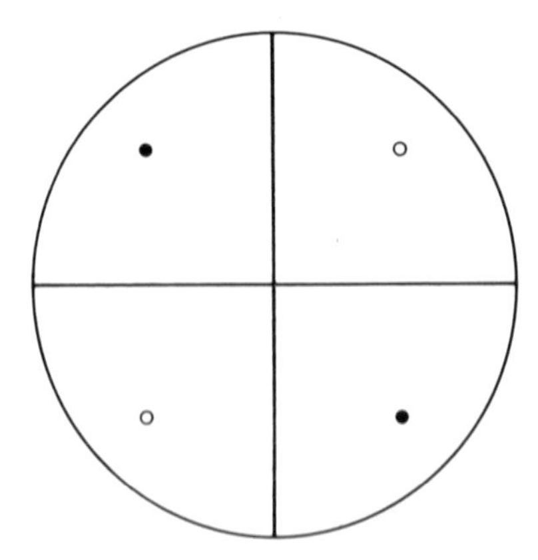

4.10 Stereogram of a sphenoid, illustrating an inversion axis.

the diagram. A rotation of 90 degrees (the normal symmetry operation of the four-fold axis) fails to produce a position of congruence, but such a position can be attained by imagining the crystal to be inverted at this stage.

Thus the four-fold inversion axis can be seen to be only a diad in the sense we have been using the term so far, but it is clearly an axis of higher symmetry than the normal diad which does not possess the inversion relationship.

Orthorhombic System

The holosymmetrical class of the orthorhombic system is represented by several important minerals characterized by the possession of three mutually perpendicular diad axes, each one lying normal to a symmetry plane. The three diads coincide with the three crystallographic axes *a*, *b*, and *c*. In this system the crystallographic axes are all of different lengths and it is desirable to have some degree of uniformity in the choice of axes. The parametral plane is usually selected, therefore, so that the *c*-axis is the shortest and the *b*-axis the longest, though other conventions have been employed.

Common forms of the orthorhombic system include:

(a) The basal pinacoid {001}, a two-faced form as in the tetragonal system.
(b) The {010} pinacoid, also conveniently referred to as the side pinacoid. A two-faced form.
(c) The {100} pinacoid, sometimes referred to as the front pinacoid. A two-faced form.
(d) A variety of prisms (i.e., four-faced forms where the faces lie parallel to the *c*-axis) such as {110} and {120}.
(e) A variety of domes (i.e., four-faced forms where the faces are parallel either to the *a*- or *b*-axes), such as {101}, {102}, {011}, and {012}.
(f) A variety of eight-faced pyramidal forms such as {111} and {112}.

Orthorhombic Minerals. Topaz (p. 138) is found in well-developed

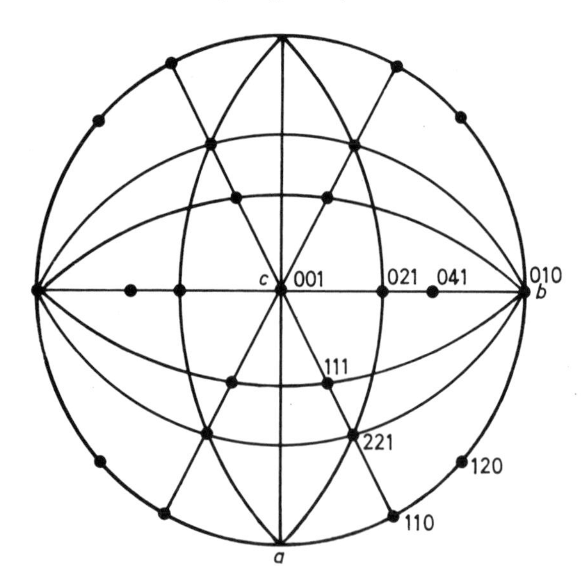

4.11 Some common forms in topaz (orthorhombic). {001} and {010} are pinacoids, {021} and {041} are domes, {111} and {221} are pyramids, {110} and {120} are prisms.

crystals and may be taken as a typical orthorhombic mineral. The mineral has axial ratios $a:b:c = 0{\cdot}53:1:0{\cdot}48$ and a stereogram of some common forms is illustrated in Fig. 4.11. A topaz crystal is shown in Fig. 1.11. Notice that correctly oriented orthorhombic stereograms are immediately distinguished from those of the cubic and tetragonal systems by the lack of the central tetrad.

Other minerals crystallizing in the orthorhombic system include *barite* (p. 116), *sulphur* (p. 101), *staurolite* (p. 142), *andalusite* (p. 138), and *olivine* (p. 133).

Hexagonal System

The essential symmetry of crystals within the hexagonal system consists of a single hexad axis. Holosymmetrical hexagonal crystals, of which *beryl* forms the only common mineralogical example, possess in addition six horizontally arranged diad axes, alternate axes coinciding with the *a*-crystallographic axes, six vertical planes of symmetry, one horizontal plane of symmetry, and a centre of

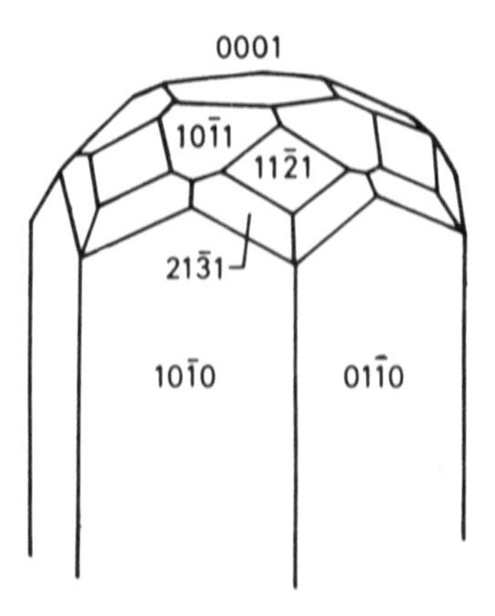

4.12 Beryl (hexagonal). Forms illustrated are the prism $\{01\bar{1}0\}$, the di-hexagonal pyramid $\{21\bar{3}1\}$, two hexagonal pyramids $\{10\bar{1}1\}$, and $\{11\bar{2}1\}$, and the basal pinacoid $\{0001\}$.

symmetry. A crystal of beryl is illustrated in Fig. 4.12 and as a stereographic projection in Fig. 4.13.

Forms commonly developed in beryl include:

(a) The basal pinacoid $\{0001\}$.
(b) Hexagonal prism of first order $\{10\bar{1}0\}$.
(c) Hexagonal prism of second order $\{11\bar{2}0\}$.

(d) Hexagonal pyramids of first order, e.g., $\{10\bar{1}1\}$.
(e) Hexagonal pyramid of the second order, e.g., $\{11\bar{2}1\}$.
(f) Dihexagonal pyramids (twenty-four-faced forms) such as $\{21\bar{3}1\}$.

The term dihexagonal is used for forms whose face-poles do not lie in symmetry planes, such forms as a result having twice the normal number of faces (cf. the use of the term ditetragonal in a previous section).

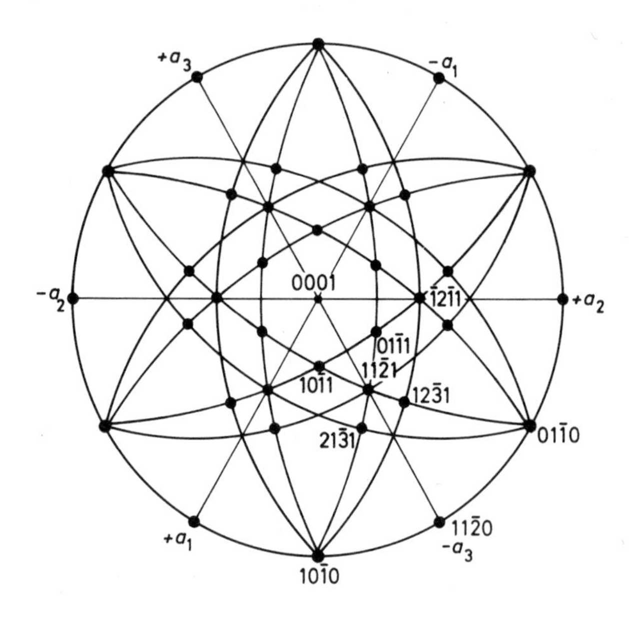

4.13 Stereogram of beryl showing the faces illustrated in Fig. 4.12 with an additional prism form $\{11\bar{2}0\}$.

Other Hexagonal Minerals. Nepheline (p. 125) and *apatite* (p. 117) are two common rock-forming minerals belonging to the hexagonal system, though both belong to classes of lower symmetry than beryl. Both have a prismatic development with the basal pinacoid prominent among the terminating faces. Nepheline crystals tend to have a stumpy habit in which the prism faces are short. Apatite in contrast, particularly when seen under the microscope in thin sections of rocks, often has an elongated acicular habit. Both minerals show hexagonal cross-sections when seen in thin section.

Trigonal System

Trigonal crystals are closely related to those of the hexagonal system and are grouped with them by some authors. In this section, several different symmetry classes within the trigonal system, respectively exemplified by the three common minerals *calcite*, (p. 113), *quartz* (p. 106), and *tourmaline* (p. 141), will be considered.

Calcite Type. Crystals of the calcite type belong to the *rhombohedral* class of the trigonal system and are typified by the cleavage rhombohedron of calcite illustrated in Fig. 4.14. The vertical symmetry axis is the triad which is the criterion of the trigonal system,

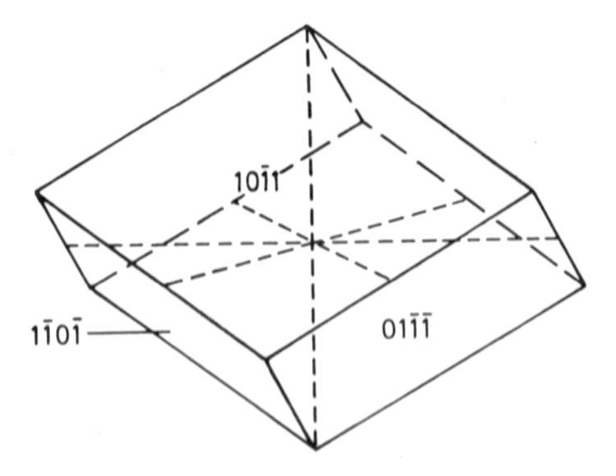

4.14 The cleavage rhomb of calcite $\{10\bar{1}1\}$ showing the positions of the crystallographic axes.

while the other axes shown (the crystallographic axes a_1, a_2, and a_3) are diads. There are three vertical planes of symmetry which lie between the diads, and there is a centre of symmetry. Notice that there is no horizontal plane of symmetry.

Calcite shows a variety of forms, the most important of which are as follows:

(a) The unit rhombohedron. This is the form taken by cleavage fragments and has the symbol $\{10\bar{1}1\}$ in the Miller–Bravais notation (see p. 15). A face of this form acts as the parametral plane.
(b) The steep rhombohedron, with the symbol $\{40\bar{4}1\}$.
(c) The flat rhombohedron $\{01\bar{1}2\}$.
(d) The scalenohedron $\{21\bar{3}1\}$.
(e) Prisms such as $\{10\bar{1}0\}$.

Natural calcite crystals show a great variety of combinations of the above forms, some of which are illustrated in Fig. 4.15.

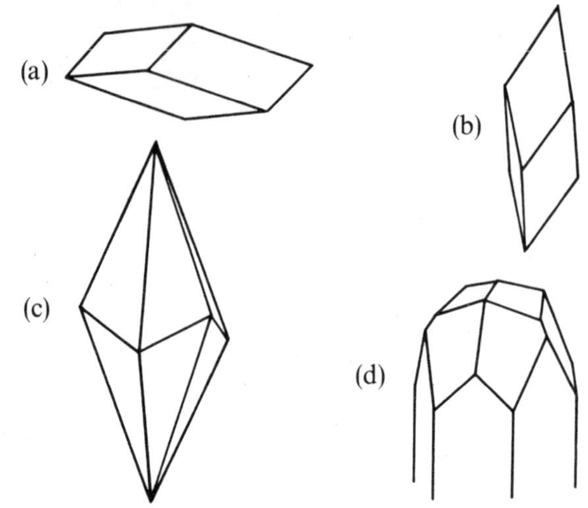

4.15 Forms of calcite. (a) flat rhomb {$01\bar{1}2$}; (b) steep rhomb {$40\bar{4}1$}; (c) scalenohedron {$21\bar{3}1$}; (d) combination of flat rhomb, scalenohedron, and prism {$10\bar{1}0$}.

A stereogram of a calcite crystal is given in Fig. 4.16.

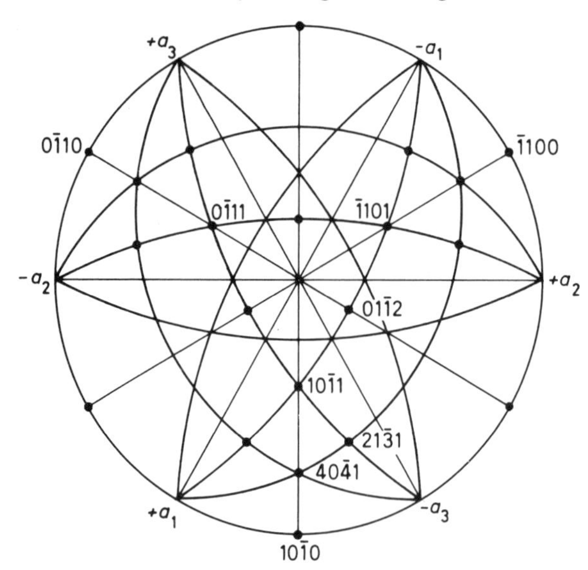

4.16 Stereogram of calcite showing forms illustrated in Fig. 4.15

Quartz. The familiar six-sided prismatic crystals of quartz, with pyramidal terminations, give the superficial impression that the mineral belongs to the hexagonal system. Firstly, however, etching of the terminal faces (e.g., with hydrofluoric acid) shows that they are of two different types and hence the main symmetry axis is a triad, not a hexad. Secondly, well-developed quartz crystals show additional faces such as the trigonal trapezohedron $(51\bar{6}1)$ which leaves no doubt as to the trigonal affinities.

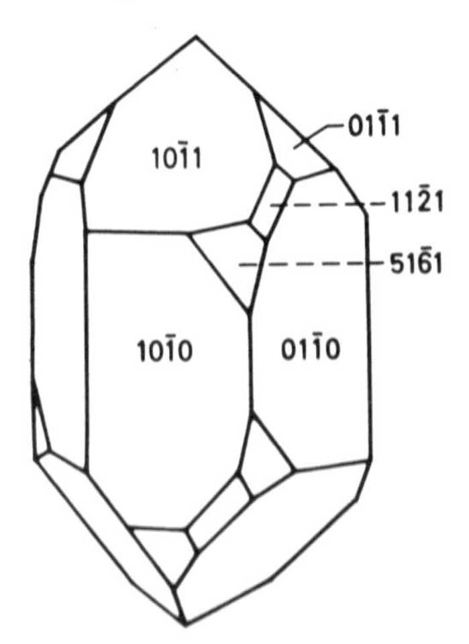

4.17 A crystal of quartz, showing the trigonal prism $\{10\bar{1}0\}$, the two rhombohedra $\{10\bar{1}1\}$ and $\{01\bar{1}1\}$, the trigonal pyramid $\{11\bar{2}1\}$, and the trigonal trapezohedron $\{51\bar{6}1\}$.

A quartz crystal showing a combination of the following forms is illustrated in Fig. 4.17, and a stereogram is given in Fig. 4.18:

(a) Rhombohedron $\{10\bar{1}1\}$.
(b) Rhombohedron $\{01\bar{1}1\}$.
(c) Prism $\{10\bar{1}0\}$.
(d) Trigonal pyramid $\{11\bar{2}1\}$.
(e) Trigonal trapezohedron $\{51\bar{6}1\}$.

It will be noted that quartz has a lower degree of symmetry than calcite since the three vertical symmetry planes are absent.

The two rhombohedra $\{10\bar{1}1\}$ and $\{01\bar{1}1\}$ are referred to as positive and negative respectively, and are analogous to the positive and negative tetrahedra discussed under the cubic system. The apparently hexagonal pyramidal terminations of quartz crystals are a combination of the two rhombohedra. Examination of natural quartz crystals will show that they are rarely equally developed.

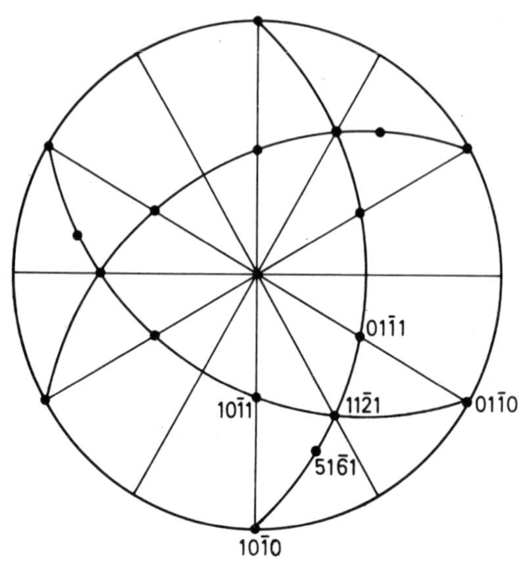

4.18 Stereogram of the quartz crystal illustrated in Fig. 4.17 (upper faces only).

Tourmaline. Tourmaline crystals (p. 141) show a feature known as *hemimorphism*, where there is no centre of symmetry and, compared with the calcite type, each form has only half as many faces. A typical tourmaline crystal is as illustrated in Fig. 4.19 where there is a very

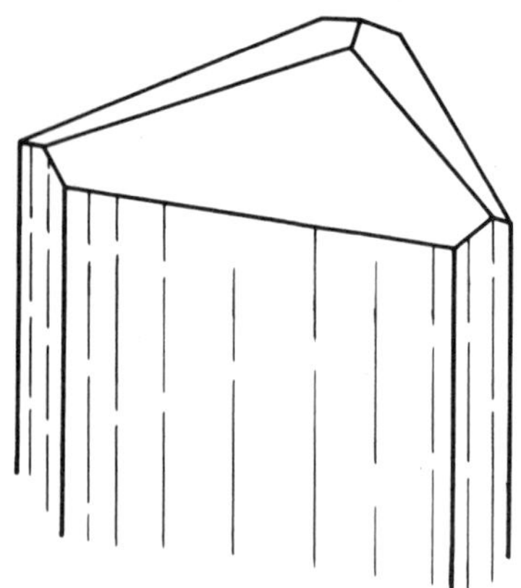

4.19 A tourmaline crystal showing unequal development of two trigonal prisms.

unequal development of the two trigonal prisms $\{10\bar{1}0\}$ and $\{01\bar{1}0\}$. The crystals are terminated by a variety of trigonal pyramids.

Other Trigonal Minerals. Amongst other trigonal minerals are the rhombohedral carbonates *dolomite* (p. 114), *siderite*, and *rhodochrosite.* These have forms similar to those of calcite. *Hematite* (p. 109) also crystallizes in the calcite class but tends to form tabular crystals in which the basal pinacoid $\{0001\}$ is prominent.

Monoclinic System

In this section we shall concern ourselves only with crystals belonging to the holosymmetrical class, since this includes several important rock-forming minerals. The essential symmetry of the class consists of a single diad axis, lying at right angles to a symmetry plane, and a centre of symmetry. By convention the diad coincides with the *b*-crystallographic axis while *a* and *c* lie in the symmetry plane (see p. 35).

It should be remembered that in monoclinic crystals the *b*-axis lies at right angles to both *a* and *c* and the obtuse angle between *a* and *c*, is termed β. The *b*-axis is known as the *ortho*-axis while *a* is known as the *clino*-axis. This nomenclature is extended to faces, as will be seen in the list of forms given below, since forms in which the faces are parallel to the *a*- and *b*-axes respectively, receive the prefixes clino- and ortho-.

Common forms of monoclinic crystals include:

(a) The orthopinacoid, $\{100\}$. Two faces.
(b) The clinopinacoid, $\{010\}$. Two faces.
(c) The basal pinacoid, $\{001\}$. Two faces.
(d) A variety of prisms, e.g., $\{110\}$ $\{210\}$. All four-faced forms.
(e) A variety of domes, those with the *a*-crystallographic axis as the zone axis, e.g., $\{011\}$ and $\{021\}$ being known as clinodomes (four-faced forms), and those with the *b*-axis as zone axis, e.g., $\{101\}$ and $\{102\}$, are two-faced forms known as hemiorthodomes.
(f) A variety of hemipyramids such as $\{111\}$. Unlike pyramidal forms in the orthorhombic system these are forms consisting of four faces only.

Monoclinic Minerals. The potassium feldspar *orthoclase* (p. 119) can be taken as a typical example of a monoclinic mineral. The crystal illustrated in Fig. 4.20 shows the unit prism $\{110\}$, the clinopinacoid $\{010\}$ and the basal pinacoid $\{001\}$, as the dominant forms. The smaller faces belong to the $\{130\}$ prism, the $\{\bar{1}11\}$ hemipyramid, and the $\{\bar{1}01\}$ and $\{\bar{2}01\}$ orthodomes.

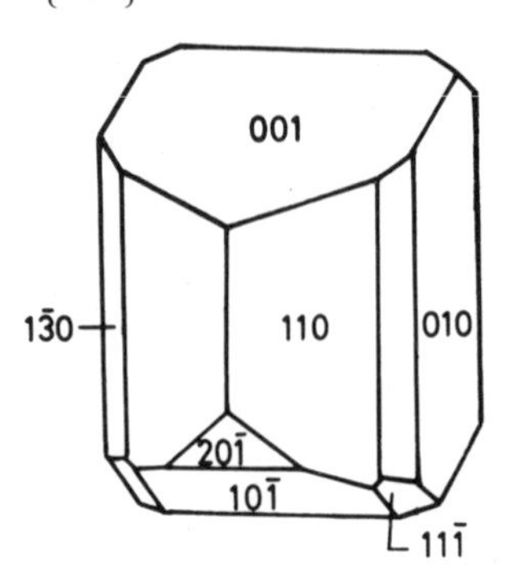

4.20 Orthoclase (monoclinic), showing the basal pinacoid $\{001\}$, the clinopinacoid $\{010\}$, and the prism $\{110\}$ as the dominant forms. Smaller faces belong to the $\{\bar{1}\bar{1}1\}$ hemipyramid, the $\{130\}$ prism, and the $\{\bar{2}01\}$ and $\{\bar{1}01\}$ orthodome forms. The upper faces of the dome and hemipyramid forms are out of sight on the back of the crystal.

A stereographic projection of the crystal is given in Fig. 4.21. The axial ratios are $a:b:c = 0{\cdot}66:1:0{\cdot}56$ and the angle β is 116 degrees.

Other monoclinic minerals include *gypsum* (p. 115), *hornblende* (p. 131), and *augite* (p. 128).

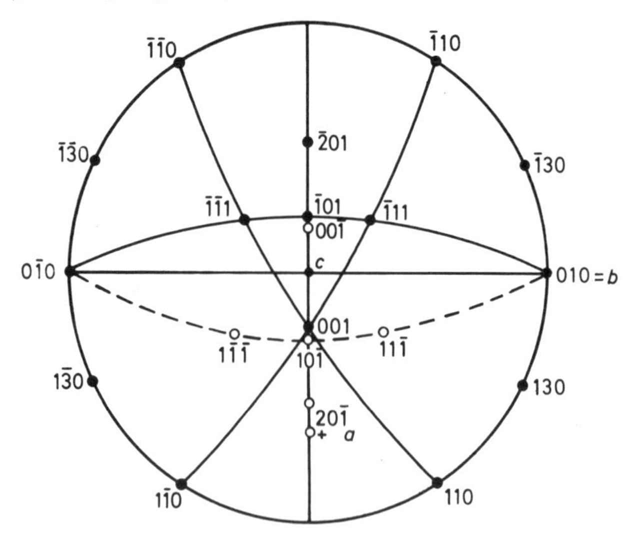

4.21 Stereogram of the orthoclase crystal illustrated in Fig. 4.20.

Triclinic System

Only brief mention will be made of this system, of which the most important representative is the plagioclase feldspar group (p. 118). The plagioclases are very similar morphologically to monoclinic feldspars such as orthoclase, covered in the preceding section. The triclinic nature is, however, shown by the inter-axial angles, which are $\alpha = 93$ degrees, $\beta = 116$ degrees, $\gamma = 88$–91 degrees. These obviously do not differ very greatly from the angles for orthoclase which may be expressed as $\alpha = 90$ degrees, $\beta = 116$ degrees, $\gamma = 90$ degrees.

Compared with the monoclinic feldspars, however, this small change in axial angles results in the loss of the diad axis and the plane of symmetry. The only remaining symmetry element is therefore the centre of symmetry, with the result that all forms developed have two faces only. The nomenclature of forms in triclinic crystals of this type is most conveniently carried out by referring to forms as pinacoidal, prismatic, etc. (depending how the faces lie relative to the crystallographic axes) and giving the form symbol in each case, e.g., the $\{010\}$ pinacoid, the $\{110\}$ prism, or the $\{\bar{1}10\}$ prism.

Twinning

When a single crystal consists of two or more parts in which the crystal lattice is differently oriented, it is said to be twinned. The geometrical relationship between the lattices in two adjacent parts of a twinned crystal is of course rational, and different possible relationships are referred to as *twin laws*. For many twins it is convenient to describe the inter-relation of the two parts by simple symmetry operations of the rotation or reflection type. Often one part of the crystal can be imagined to fall into congruence with the other part by a rotation of 180 degrees about a line known as the *twin axis*. The twin axis is always parallel to a possible edge or perpendicular to a possible face. In many cases the same resulting position of congruence can be obtained by imagining reflection to take place across a plane known as the *twin plane*. A simple example of this is afforded by the common twinned crystals of gypsum known as swallow-tail twins. Here the twin plane is the orthopinacoid (100) and the twin axis is at right angles to this (see Fig. 4.22). This gives an example of a *contact twin* in which the two units of the twin come into contact in a plane known as the *composition plane*. The composition plane and the twin plane often coincide but do not necessarily do so.

Not uncommonly, twins penetrate each other, often in a somewhat irregular fashion and there is no composition plane as such.

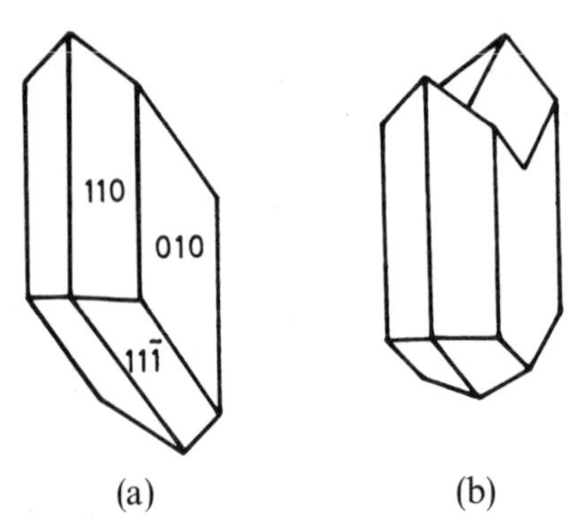

4.22 Twinning in gypsum (monoclinic). (a) untwinned crystal showing the clinopinacoid {010}, prism {110} and hemipyramid {$\bar{1}$11}. (b) a 'swallow-tail' twin of gypsum with (100) as twin plane.

Repeated twinning (*polysynthetic twinning*) is a prominent feature of many minerals, particularly the plagioclase feldspars. In the latter crystals, twinning with (010) as twin and composition plane, known as the *albite* law, is repeated on a very fine scale, and divides the crystals up into narrow lamellae with alternate orientations. This is a very obvious feature of plagioclase under the microscope (see

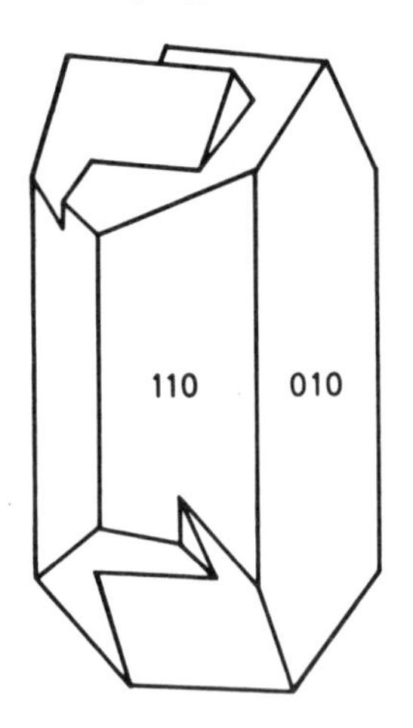

4.23 A Carlsbad twin of orthoclase (monoclinic). The twin plane is (010) and the twin axis is the *c*-crystallographic axis.

Fig. 7.11) but may also be observed as fine striations on the basal pinacoid in hand-specimens.

Feldspars in general also commonly show twinning according to the *Carlsbad* law in which the twin axis is the *c*-crystallographic axis and the composition plane is (010). Interpenetrating twins of this type are common and are illustrated in Figs. 4.23 and Fig. 7.10.

Other minerals in which simple twinning is frequently encountered include the pyroxenes and amphiboles where the twin plane is (100). Fluorite and staurolite show interpenetrating twins illustrated in Fig. 4.24, while rutile forms twins which are termed geniculate. In

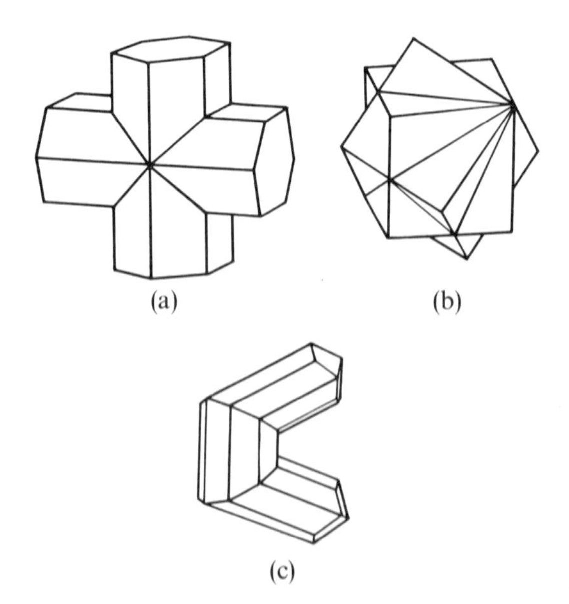

4.24 (a) Interpenetrating twin of staurolite (orthorhombic); (b) interpenetrating twin of fluorite (cubic); (c) geniculate twin of rutile (tetragonal).

many of the twinned crystals illustrated it will be observed that re-entrant angles occur. This is not present with all twin laws but is sufficiently common to act as a general characteristic of twinned crystals in hand-specimens.

5. Macroscopic characters of minerals

Introduction

Minerals are naturally occurring substances, usually crystalline, whose compositions are either fixed or can only vary between certain fixed limits. Rocks, in contrast, are aggregates of mineral grains, usually belonging to more than one mineral species. Rarely, one may find monomineralic aggregates, e.g., the igneous rock dunite, composed entirely of olivine, but these are best referred to as monomineralic rocks.

Some minerals have a constant chemical composition and are definite chemical compounds, e.g., quartz (SiO_2), fluorite (CaF_2), and barite ($BaSO_4$). Others, indeed the majority of the rock-forming minerals, are solid-solution series between two or more compounds, termed end-members, of fixed composition. Thus the important group of plagioclase feldspars ranges in composition from the one end-member albite ($NaAlSi_3O_8$) to the other, anorthite ($CaAl_2Si_2O_8$) via all possible intermediate compositions as calcium replaces sodium, and aluminium replaces silicon in the lattice. Throughout such a series there is a gradual change of physical properties which is, however, rarely noticeable in the hand-specimen, the study of such changes being best carried out by the methods of microscopy, some of which are described in chapter 6.

Characters of Minerals in Hand-Specimen

Minerals in hand-specimens can be identified by an examination of their crystal form, hardness, specific gravity, colour, lustre, degree of transparency, streak (colour of the powder when crushed), and occasionally by their taste, smell, or feel.

It is rarely necessary or even possible to assess all of these characters in a given specimen. The art of rapid mineral identification lies in using certain selected properties to distinguish between the various possibilities that a preliminary examination immediately suggests. For example, an opaque golden-coloured mineral with a metallic lustre is likely to be either pyrite (FeS_2) or chalcopyrite ($CuFeS_2$), and these two may easily be distinguished by their hardness. Similarly barite may superficially resemble a number of other common minerals including calcite and fluorite. The disposition of the cleavages will distinguish the three, but in addition barite has a very much higher specific gravity than the other two. Other properties such as colour and streak would be of no determinative value in this particular case. With experience it becomes possible to recognize many common minerals at a glance by an almost sub-conscious appreciation of their properties. The ability to describe a given specimen fully is, however, essential, because it is otherwise impossible to identify a mineral which has not previously been met. This point, that is, the idea that a mineral is identified by an assessment of all its individual characters, cannot be overstressed. Simply learning to recognize the specimens in a given collection is of no value whatever.

Crystal Form

In well crystallized specimens the determination of crystal system can be a valuable aid to identification. Irregular development of faces is very common so that it is often essential to look at interfacial angles, rather than overall shape, in determining the symmetry elements. The habit of the crystals, i.e., the way in which the various forms are developed, and the forms of crystalline aggregates, while rarely being diagnostic can nevertheless be a useful guide to identification in preliminary examinations. Some terms used to describe habit and form are listed below:

Acicular—fine needle-like crystals, e.g., zeolites.
Bladed—elongated crystals flattened in one direction, e.g., kyanite.
Botryoidal—rounded masses somewhat resembling bunches of grapes, e.g., chalcedony.
Fibrous—groups of parallel thread-like strands, e.g., zeolites, asbestos.
Mammilated—rounded masses similar to the botryoidal form but the protuberances are more flattened, e.g., malachite.
Massive—crystalline aggregates with no regular form.

Micaceous—splitting readily into thin plates, e.g., muscovite, biotite.

Platy—crystals extremely flattened in one direction.

Prismatic—somewhat elongated crystals with well-developed prism faces, e.g., quartz, hornblende.

Reniform—rounded kidney-shaped masses, e.g., hematite.

Tabular—crystals somewhat flattened in one direction, e.g., commonly barite, feldspars.

Filiform—wire-like, twisted crystals, e.g., native silver.

Cleavage

Many minerals split along well-defined planes termed cleavage planes which may or may not be parallel to faces but which are related to the internal structure of the crystal just as faces are. Hence on the basis of cleavage alone it is often possible to determine the symmetry of a crystal and thus its crystal system. This is one of the most useful determinative techniques because crystals showing a good cleavage are more common than crystals showing well-developed faces.

Cleavage may be studied by breaking the crystal or more often simply by examining it. Cleavage planes can be seen intersecting the faces of the crystal, and, if the crystal is transparent, running through its interior. It is important to determine the number of different cleavage directions present and their approximate angular relations to each other, in order to extract the necessary information about the symmetry of the crystal.

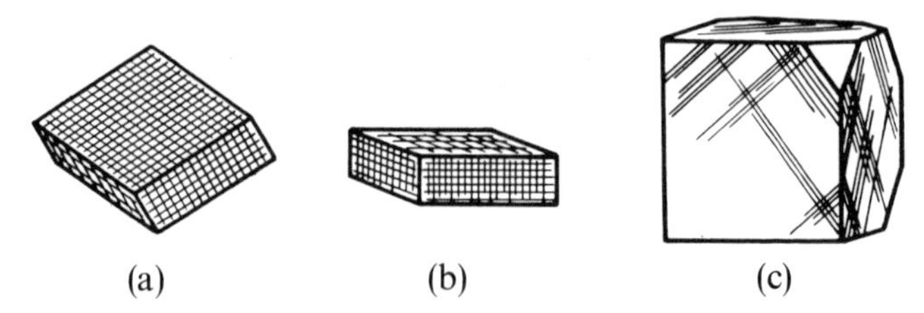

5.1 Cleavage of minerals. (a) calcite (trigonal). Three cleavage directions parallel to the faces of the $\{10\bar{1}1\}$ rhombohedron; (b) barite (orthorhombic). Three cleavage directions. One is parallel to the basal pinacoid $\{001\}$ and the other two are parallel to (110) and $(\bar{1}10)$. These lie at right angles to the basal cleavage and at 78 degrees to each other; (c) fluorite (cubic). Four cleavage directions parallel to the faces of the octahedron. Cubes of fluorite are often seen minus their corners as a result of the excellence of this cleavage. The cleavage surfaces have a conspicuously different lustre from that of the crystal faces.

Figure 5.1 illustrates the cleavage of the three common minerals calcite (trigonal), barite (orthorhombic), and fluorite (cubic). In each case a sufficiently careful examination of the cleavage alone is sufficient to reveal the crystal system. Cleavage rhombs of calcite show for example a solitary triad axis, cleavage fragments of barite show three diad axes (and three symmetry planes), while the octahedral cleavage fragments of fluorite show four triad axes. It is of course more difficult to determine the symmetry by the method of observing the intersections of cleavage planes with the crystal faces. This method is, however, the only one available when the mineral concerned is too hard to break into cleavage fragments or when it is desirable not to damage the specimen.

Fracture

Some minerals instead of cleaving neatly when broken, fracture in a more or less irregular fashion. Occasionally the appearance of the fractured surface may have some slight diagnostic value. The curving fracture marked with concentric arcuate ridges, which is characteristic of broken glass, for example, is also characteristic of quartz and is known as a *conchoidal* fracture because of its resemblance to certain types of sea-shell. Metals also have a characteristic fracture known as *hackly*, in which the fractured surface is covered with small, jagged points. This is characteristic of the mineral native copper.

Hardness

The hardness of a mineral, because it can be expressed in a semi-quantitative fashion, is an extremely useful property to determine in mineral identification. Hardness is measured on the Mohs Scale (see Table 7.2), each mineral in the scale being hard enough to scratch the one below it and soft enough to be scratched by the one above it. Thus a set of mineral specimens corresponding with the Mohs Scale can be assembled and used to test unknown specimens, though it should be remembered that the scale arranges minerals only in *order* of hardness and the numbers in the scale do not express the actual hardness of a mineral in any exact fashion. The testing procedure must be carried out with some care, preferably using a hand lens to examine the scratches produced on one specimen by the other, since it often happens that the softer mineral leaves a trail of crushed powder on the harder one and this may superficially resemble a scratch.

Rapid and approximate hardness testing is also very useful and employs such common objects as the finger nail, the copper coin, and the knife blade. These have approximate hardnesses on Mohs Scale of 2, 3, and 5 respectively, and may often be used to distinguish different possibilities suggested by a preliminary examination. If we take for example some common white or colourless minerals such as quartz, calcite, and gypsum we can easily distinguish these since quartz (hardness 6) cannot be scratched by the knife, calcite (hardness 3) can be scratched by the knife but not by the finger nail, and gypsum (hardness 2) can be scratched by both. With experience it is possible to put an approximate figure on the hardness if it is 5 or less by seeing how easy or difficult it is to scratch the mineral with the knife.

Specific Gravity

The specific gravity of a mineral is rarely determined exactly during routine mineral identification procedures but may usually be determined very roughly by weighing the specimen in the hand. Since most rocks and minerals have specific gravities in the range 2·5–3·0 the mind tends to have a built-in preconceived idea of how heavy a specimen ought to feel when weighed in the hand, irrespective of its size. This is presumably based on many years of experience of throwing stones, etc. Thus any specimen whose density departs notably from this range will feel abnormally heavy or light. Most departures from normality are on the heavy side, particularly amongst the common ore minerals such as galena (PbS, density 7·5), magnetite (Fe_3O_4, density 5·2), pyrite (FeS_2, density 5·0), chalcopyrite ($CuFeS_2$, density 4·2), cassiterite (SnO_2, density 6·9). Amongst other common minerals, barite with a density of 4·5 is notable since in superficial appearance it is rather like calcite, fluorite, quartz, etc., all of which are minerals having 'normal' densities. It is thus a particularly easy mineral to identify.

Colour

The colour of mineral specimens is unfortunately rarely of much use in mineral identification since it can be greatly influenced, particularly in transparent or translucent minerals, by small amounts of impurities. Thus a mineral such as barite, while commonly pale in colour or almost white, can often be found in pinkish, greenish, bluish, or yellowish specimens. Quartz also is a good example of this type of behaviour and occurs in many varieties such as rock crystal (colourless), amethyst (violet), rose quartz (pink), smoky quartz (grey), and cairngorm (whisky-coloured). Fluorite similarly is found in a variety of colours, of which violets, blues, greens, and yellows are perhaps the most common.

Certain minerals, however, while also being somewhat variable, nevertheless have characteristic colours which are an aid to recognition. Many of the opaque minerals fall in this category, particularly malachite (green), pyrite and chalcopyrite (golden), galena (silvery grey). Other minerals such as cinnabar (HgS, bright red), realgar (AsS, orange-red), native sulphur (yellow), azurite (deep blue), and epidote (pistacchio green) are also distinctive.

Streak

Often the colour of the crushed powder of a mineral specimen is of more determinative value than the colour of the mineral itself. This is known as the *streak* of the mineral, and is observed by rubbing the specimen across a white unglazed porcelain plate known as a streak plate. The powder adheres to the plate and its colour is easily assessed. The colour of the powder can also be observed by scratching the specimen. Most pale-coloured, transparent to translucent minerals have a white streak which is therefore of no determinative value, but the method is of great use with the many opaque minerals which are frequently dark or blackish in colour. Hematite, for example, can appear as shiny black crystals (specular hematite) or as dull reddish masses (kidney iron ore) of totally different appearance. Both of these, however, give the red streak characteristic of the mineral. Similarly, limonite which can occur in a variety of earthy, brownish or botryoidal black masses has a characteristic yellow ochre streak. Other minerals with distinctive streaks include wolfram (chocolate brown), cinnabar (scarlet), and pyrite and chalcopyrite (greenish black).

Lustre

The lustre observed on crystal faces or on fresh cleavage surfaces can be of considerable importance in mineral identification though many of the differences are somewhat subtle and some experience is necessary before full use can be made of this property. Terms used to describe lustre are listed below with examples:

Metallic—galena, pyrite, chalcopyrite.
Pearly—selenite (gypsum), some varieties of barite.
Resinous—sphalerite (zinc blende).

Silky—shown by some fibrous minerals such as the variety of gypsum known as satin spar.

Vitreous—the lustre of broken glass. Quartz. Calcite has what may be termed a sub-vitreous lustre, as have many pale-coloured minerals.

In addition to these terms describing the quality of the lustre several others are used to describe the intensity of the reflection from a surface. These include *splendent*, when the reflection is intense enough to give a mirror-like effect (e.g., specular hematite), and such terms as shining, glistening, and dull to describe lesser degrees of reflection.

Degree of Transparency

The terms transparent, translucent, and opaque are commonly used to describe the light-transmitting properties of minerals in hand-specimens. In any individual specimen, however, this depends very much on the thickness, and in fact any mineral will transmit some light in thin enough sections. The degree of internal alteration, the presence of inclusions, and the extent to which cleavage planes are developed inside the specimen are also important in determining the transparency. Thus, as a mineral property, the degree of transparency is only of limited use in mineral identification. Many minerals are notably transparent when well crystallized (e.g., quartz, calcite, fluorite, barite) but are often seen as translucent specimens. Other minerals such as the feldspars are rarely more than translucent, and others including the majority of the ore minerals (e.g., magnetite, pyrite, galena) are for practical purposes always opaque.

6. Minerals in thin section

Introduction

A thin section is made by grinding down a slice of rock which has been gummed to a glass slide, until it reaches a standard thickness of 0·03 mm. At this thickness almost all minerals become more or less transparent and they can therefore be studied by microscopy using transmitted light. The advantages of this system, compared with the hand-specimen method, are enormous, and the identification of common minerals becomes relatively easy even when the rock under examination is fine-grained. The instrument used in these studies is the polarizing microscope, in which the light is polarized before transmission through the crystal. Many properties of crystalline substances which are not apparent in ordinary (non-polarized) light may then be observed.

The treatment of the theory of optical mineralogy given in this chapter is restricted, and does not include the subjects of interference figures and compensating plates. The great majority of rock-forming minerals can, however, be identified without these advanced techniques, although reference will be made to them occasionally for the benefit of students who may be proceeding to more advanced courses.

The Nature of Polarized Light

Light travels as electro-magnetic vibrations in which the vibration direction is transverse to the direction of propagation. Transverse wave-motions of this type are said to be *plane-polarized* when all the vibrations lie in one plane. This contrasts with the ordinary or non-polarized condition in which vibrations in all possible directions normal to the direction of propagation are simultaneously present.

The Polarizing Microscope

The normal non-polarizing microscope consists essentially of a light source, a sub-stage condenser, a stage to hold the specimen, an objective, and an eye-piece. The simple polarizing microscope carries in addition a device for producing polarized light, termed the polarizer, a graduated rotating stage as opposed to a fixed stage, and a second polarizing device termed the analyser.

The *analyser* and *polarizer* are mounted sheets of Polaroid, a substance which allows the passage of light with only one vibration direction. The polarizer is usually mounted below the stage, and the analyser in the microscope tube where it can be slid into or out of position as required. The two devices are arranged so that the vibrations they transmit are mutually perpendicular. Before the invention of Polaroid, calcite crystals were used to make polarizers termed Nicol prisms, and the name persists. With the analyser in position observations are said to be made 'between crossed nicols'; with the analyser out of position observations are referred to as being 'in plane polarized light'.

The *sub-stage condenser* is an important part of the microscope if medium- or high-power objectives are to be used. On most models rack-and-pinion focussing is provided, and the position of the condenser has a considerable effect on the resolution of the object viewed, and also on the evenness of the illumination. The microscope usually also incorporates a sub-stage iris *diaphragm* which in use effectively alters the apparent contrast of the image.

Finally the simple polarizing microscope incorporates *cross-wires* in the eye-piece, which are parallel to the vibration directions of the analyser and polarizer. Depending on the microscope used, the polarizer vibration direction may lie parallel to either cross-wire.

Double Refraction

Most crystalline substances are *anisotropic*, that is to say their physical properties differ if measured in different directions. Crystals belonging to the cubic system are the exception and are said to be *isotropic*.

When a ray of ordinary light enters an anisotropic crystal it is in general split into two rays which are plane polarized in mutually perpendicular planes and travel through the crystal in slightly different directions. This is the phenomenon known as 'double refraction' and may readily be observed by placing a cleavage rhomb of calcite over a small spot marked on a piece of paper. A double image is seen if the spot is viewed through the calcite.

The polarization of the two images, representing the two rays, may be tested by the use of a Polaroid sheet. A polarized ray will pass unaffected through the Polaroid, providing the vibration direction of the ray coincides with the vibration direction of the Polaroid. Conversely, the ray will be completely eliminated if its vibration direction is at right angles to that of the Polaroid. At intermediate angles a proportion of the light will be transmitted and the remainder absorbed. Thus if one of the spots produced by the calcite rhomb is observed through a piece of slowly rotated Polaroid, a position will be found where the image disappears completely. Rotation of the Polaroid through a further 90 degrees will then cause the other spot to disappear while the first reappears.

The two rays, apart from being polarized at right angles to each other, also travel at different speeds through the crystal, i.e., they have different refractive indices.* This is again easily demonstrated with the calcite rhomb, since if the observer moves his or her head from side to side it will be seen that the two images are at different apparent depths below the surface of the calcite.

The apparent depth of an object viewed through a layer of transparent material is a function of refractive index, given by the expression:

$$\text{Apparent depth} = \frac{\text{Real depth}}{\text{Refractive index}}$$

It may be recalled that this forms the basis of an elementary method of determining refractive indices of liquids, using a travelling microscope to measure object and image positions.

It should be emphasized that the phenomenon of double refraction, that is the production of two rays with mutually perpendicular vibration directions and different refractive indices, is not simply a property of calcite but of *all* anisotropic crystals. Calcite, however, is one of the best substances to demonstrate it with because it is easily obtained in clear crystals and produces a greater separation between the two rays than most other minerals.

Determination of Vibration Directions in Mineral Grains in Thin Sections

Light transmitted through an anisotropic mineral grain as seen in thin section with the polarizing microscope travels, in general, as two rays with mutually perpendicular vibration directions. The positions of the vibration directions are fixed with respect to the

* Refractive index is defined as the ratio of the speed of light *in vacuo* to its speed in the substance.

grain and hence rotation of the specimen on the stage results in the rotation of the directions. The light illuminating the specimen is, however, already plane polarized and in certain positions of the specimen the plane of polarization of the illuminating beam will fall into parallelism with one or other of the vibration directions within the crystal. In this situation all the light passing through the crystal utilizes one of the vibration directions only since it has no component of vibration in the plane of the other.

If, on the other hand, the plane of polarization of the illumination is not parallel to either of the vibration directions of the grain, light will pass through the crystal utilizing both of the possible vibration directions (see Fig. 6.1).

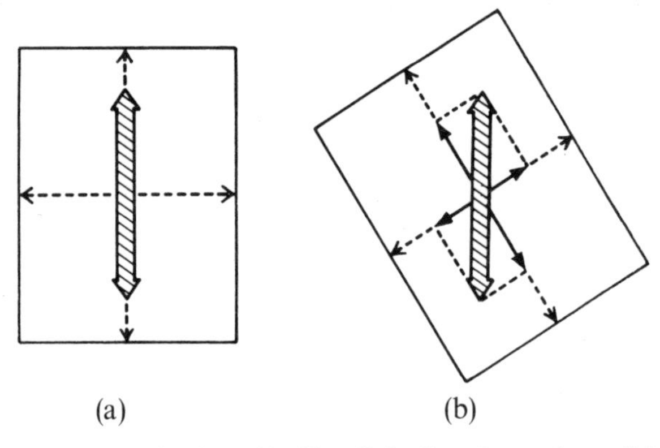

6.1 (a) Plane of polarization of incident light (broad arrow) parallel to one of the vibration directions of the crystal (broken arrows). All the light is transmitted utilizing the north–south vibration direction only as there is no component of the incident light in the east–west direction.

(b) Plane of polarization not parallel to the vibration directions of the crystal. Light is transmitted as two components (black arrows).

Let us now consider the function of the analyser, a device like the polarizer but lying above the specimen in the microscope tube and arranged so that its vibration direction is at right angles to that of the polarizer. Inserting the analyser into position will, in the absence of a specimen on the stage, result in a dark field of view because polarized light reaching the analyser has no component of vibration in the vibration direction of the analyser and is hence completely absorbed.

If an anisotropic crystal is now placed on the stage it will, in general, appear illuminated and coloured. The colours seen are known as *polarization colours* and will be considered in a later section. Note, however, that the mineral grain will become *dark* in four positions 90 degrees apart as the stage is rotated through a complete revolution. This phenomenon is known as *extinction* and the settings of the stage necessary to produce darkness are known as *extinction positions*.

Extinction occurs whenever either of the vibration directions of the grain falls into parallelism with the vibration direction of the polarizer. In this position, as we have seen, all the light transmitted through the crystal utilizes one vibration direction only, that which

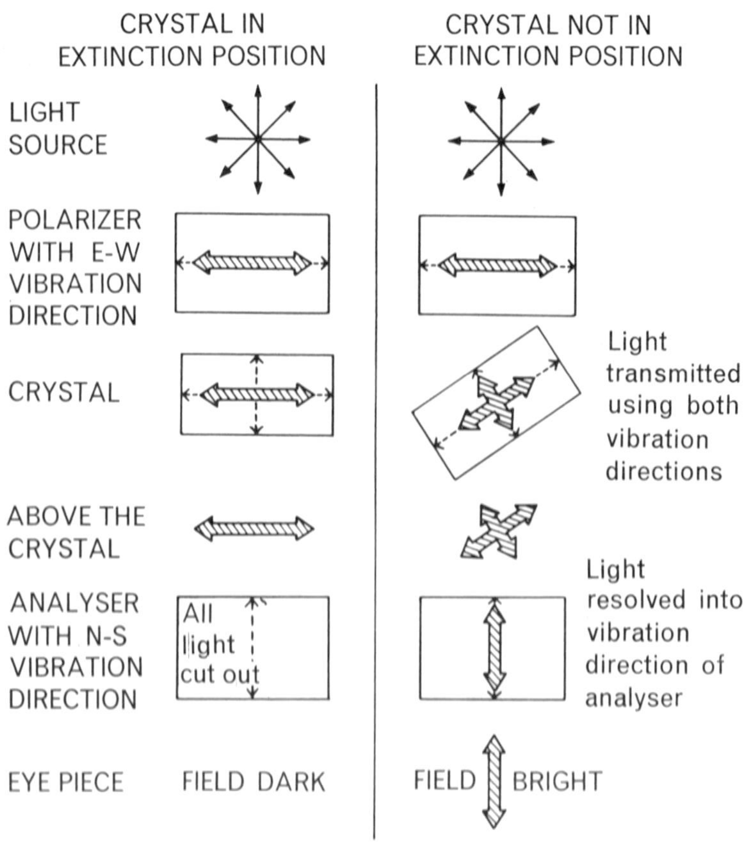

6.2 Each figure represents a section cut across the optical path within the microscope at the points indicated on the left. Vibration directions of the crystal indicated by broken arrows, light transmitted by broad shaded arrows.

is parallel to the polarizer vibration direction. This light must therefore be completely cut out by the analyser just as it is when no specimen is present. At settings of the stage in between the four extinction positions, some light will always pass through the analyser because the light passing through the crystal utilizes both vibration directions, neither of which is normal to the vibration direction of the analyser (see Fig. 6.2).

Thus the determination of extinction positions is used as a means of locating vibration directions in a mineral grain. This is a particularly useful technique because, amongst other things, it makes possible the study of the two rays *one at a time*, an essential requirement in the determination of pleochroism and relief, two properties discussed in the next section.

Properties of Minerals in Plane-polarized Light

(a) *Crystal Morphology.* Before discussing pleochroism and relief, both of which depend for their study on the provision of a source of polarized light, it is convenient to consider the morphology of crystals as they appear in thin sections. This property can be studied on a non-polarizing microscope since it is in no way dependent on the nature of the illumination. It is, however, most easily studied on a petrological microscope where the rotating stage makes possible the measurement of angular relationships which could otherwise only be estimated. The sub-stage diaphragm is used throughout the observation of morphological features to obtain the necessary image contrast.

The forms of crystals and the arrangement of cleavages within them are useful determinative characters. The microscopist must, however, be able to relate the different two-dimensional sections he sees in such a way as to acquire a three-dimensional impression of the mineral under observation. To this extent, the observation of crystal form is more difficult under the microscope than it is in the hand-specimen, but this is outweighed by the ease with which cleavages may be observed and their angular relations established.

Angular relations between cleavages can be measured provided that the cleavages concerned are normal to or nearly normal to the plane of the slide. One of the cleavage traces is set parallel to one of the cross-wires and the reading on the stage scale is noted. The stage is turned until the second cleavage becomes parallel to the same cross-wire and a second reading of the scale is made. The angle between the two cleavage traces is then the difference between the two scale readings.

Measurement of angles between cleavages can be a useful identificatory technique, particularly in the distinction between amphiboles (two cleavages at 124 degrees) and pyroxenes (two cleavages at nearly 90 degrees) (see Fig. 6.3).

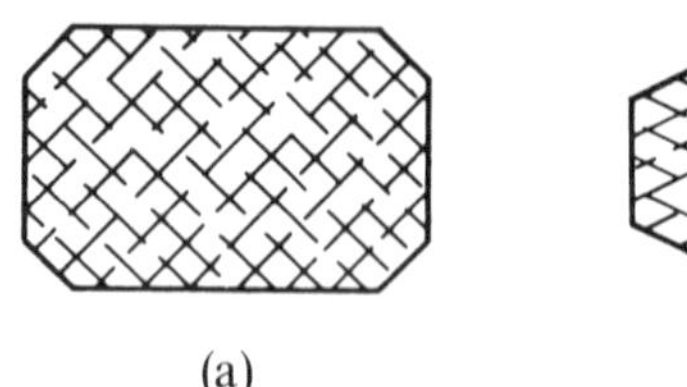

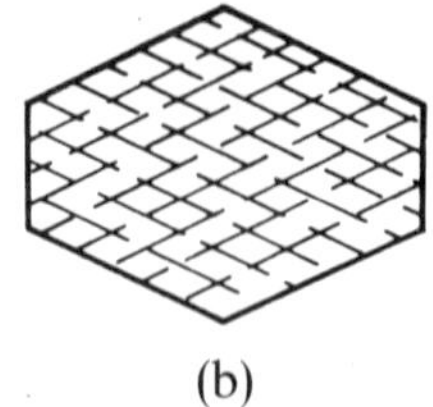

(a) (b)

6.3 (a) Section normal to *c*-axis in augite, showing the two cleavages of the {110} form nearly at right angles to each other; (b) a similar section of hornblende showing the {110} cleavages at 124 degrees.

The study of crystal form presents problems similar to those of the study of cleavages, problems largely due to the difference between the individual two-dimensional sections seen and the true three-dimensional form of the grains. Minerals which, for example, appear in the thin section as elongated grains may genuinely have an elongated crystal form or they may merely be flattened in one direction. It is possible to distinguish these two cases by a careful study of numerous grains within the same thin section. In Fig. 6.4 the contrast is shown between well-formed tabular grains of biotite and

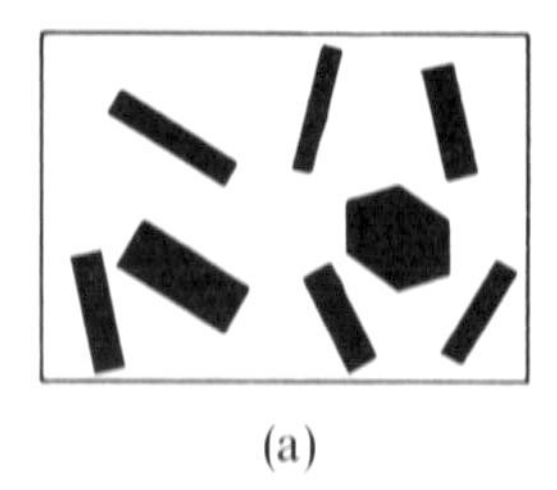

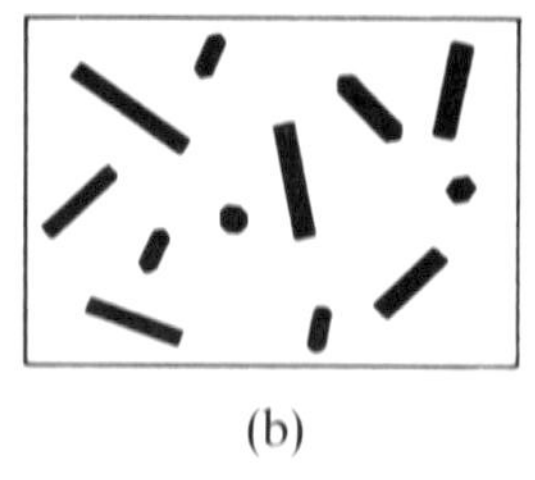

(a) (b)

6.4 Shape in thin section of (a) tabular biotite contrasted with (b) prismatic apatite.

well-formed prismatic grains of apatite as they appear in thin section. In the case of the tabular crystals, a few equidimensional sections are seen (sections cut parallel to the plane of flattening of the crystals) and these have the same diameter as the length of the elongated sections. In the prismatic case, the equidimensional grains have a diameter only equal to the width of the elongated grains. In both cases note that elongated or mildly elongated grains are much more common than equidimensional grains.

With these considerations in mind, the study of crystal shapes can be extremely helpful in mineral identification since a number of common rock-forming minerals often exhibit characteristic forms. Well-formed pyroxenes, for example, often appear as stumpy prismatic crystals with a characteristic eight-sided cross-section bounded by the {010}, {100}, and {110} forms (see Fig. 6.3). Amphiboles in contrast often give six-sided cross-sections bounded by the {010} and {110} forms only. The tabular form is characteristic of all the micaceous minerals and is often discernible even if the grains are not well formed. An almost circular section is characteristic of certain cubic minerals such as leucite and garnet (see Figs. 7.17 and 7.26). Olivine crystals when well formed, e.g., in volcanic rocks, are also distinctive (see Fig. 7.23).

(b) *Relief*. It is immediately noticed, on examining any slide containing a reasonable selection of different minerals, that some are clearly visible, that is, details of surface texture, cleavage, etc., are obvious, while others appear almost completely featureless and, if colourless, barely visible (see Fig. 6.5).

We are here observing the property known as 'relief'. Minerals which have refractive indices which differ markedly from that of the mounting medium show up clearly in thin sections and are said to have high relief. Minerals with low relief conversely have refractive indices close to that of the mounting medium, which is usually Canada Balsam with a refractive index of 1·54. Relief is spoken of as positive when the mineral has refractive indices above the balsam, and negative when its indices are lower than that of balsam. Thus olivine, with refractive indices in the range 1·64–1·88 has a high positive relief, quartz with refractive indices of about 1·55 has a low positive relief, orthoclase with refractive indices of about 1·53 has a low negative relief, and leucite with an index of 1·51 has a moderate negative relief.

The determination of the sign of the relief is easily made using the *Becke test*. It will be observed that where a mineral grain is adjacent to the Canada Balsam, e.g., at the edge of the slide or where the grain is cracked, that a bright line of light, the Becke line, can be seen marking the boundary between the balsam and the grain (see Fig. 6.6). This is best observed using a low- or medium-power objective with the sub-stage diaphragm almost closed. As the microscope tube is racked slightly up or down it will be observed that the Becke line moves sideways. The relief of the grain is then easily determined from the following rule: as the tube is racked *down* the Becke line moves towards the medium with the *lower* refractive index. Hence it may readily be ascertained whether the mineral or the balsam has the lower index. The procedure may be applied equally well to two adjacent minerals if further information is needed.

6.5 Illustration of relief in thin sections. The diamond shaped section of sphene shows very high relief. Beside it biotite shows high relief and a conspicuous cleavage while the remainder of the photograph is occupied by quartz and feldspar, both of which have low relief. × 100. Plane polarized light.

For a more rigorous treatment of the Becke test it is, however, necessary to remember that the light passes through the crystal as two rays, each of which has its own refractive index. The relief observed is normally therefore a compound effect related to both rays. If the nicols are crossed, however, and the crystal is rotated to an extinction position before the relief is examined (after uncrossing the nicols), one can be certain that the relief observed is that corresponding to one ray only. Rotation of the specimen through 90 degrees will then bring the other ray into use in its turn.

Normally the two reliefs observed by this method are not notably different, but in specimens of very low relief it not infrequently happens that the relief is of opposite sign for the two rays, that is to say the refractive index of one ray is slightly above that of balsam while the other is slightly below. A few common minerals do, however, show a considerable change of relief when rotated from one extinction position to the other, a phenomenon known as *twinkling*.

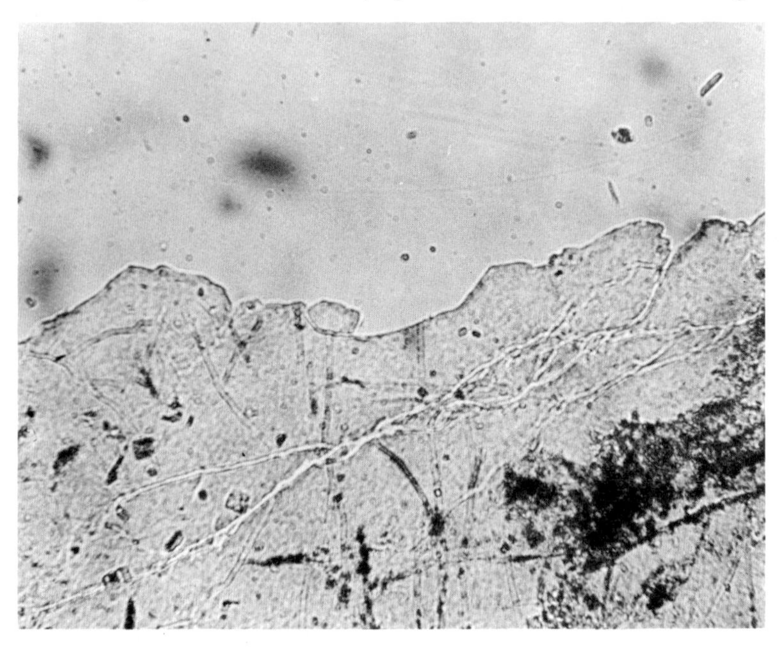

6.6 The Becke line. The photograph shows feldspar (moderate relief) in contact with Canada Balsam (no relief). The Becke line is seen as a bright line of light along the contact. × 100. Plane polarized light with sub-stage diaphragm closed down.

Sections of muscovite, for example, cut so that the basal cleavage is normal to the plane of the slide, transmit rays with refractive indices of about 1·60 and 1·56. The first of these corresponds with a fairly high positive relief and the second with a rather low positive relief. The so-called twinkling effect is obtained by rotating the stage rapidly. Calcite shows an even more marked change of relief, in sections cut parallel or nearly parallel to the *c*-axis, where the two rays have refractive indices of 1·49 and 1·66 respectively (cf. Canada Balsam 1·54).

(c) *Pleochroism*. The colour observed in a mineral grain in thin section is the result of selective absorption of certain of the wavelengths which make up the white light supplied by the light source. Not surprisingly, the anisotropism shown by crystals in their other physical properties is also shown by their absorption, since the wavelengths absorbed and also the amount of light absorbed depend on the vibration direction of the light transmitted. This is the phenomenon known as *pleochroism* and is apparent in thin sections when coloured minerals undergo a colour change as they are rotated in plane polarized light. In the previous discussion of relief it was noted that the two rays in a given grain could be studied separately, so far as their effects on the relief of the grain were concerned, if the mineral was turned successively to its two extinction positions between crossed nicols, before examination. The absorption effects of the two rays may be studied in exactly the same way but it must be remembered that just as one given section (grain) may show different colours in its two extinction positions, so two different grains of the same mineral, if they are cut with different orientations relative to the crystal structure, can show different effects. Thus crystals of biotite cut normal to the basal cleavage commonly show a very marked colour-change from dark brown to pale brown as they are rotated in plane polarized light. In contrast, sections cut parallel to the cleavage show virtually no pleochroism and are dark brown in all positions. Sections cut at intermediate angles show an intermediate type of pleochroism, from moderately dark to moderately pale brown.

Apart from biotite, hornblende is perhaps the most common strongly pleochroic mineral and is sometimes noticeably *trichroic*, that is to say, in contrast with biotite, which shows pleochroism based on two colours only and may be referred to as *dichroic*, hornblende has an absorption scheme based on three colours. Different sections of the same hornblende can therefore show, for example, pleochroism from neutral to dull green, dull green to blue green, and blue green to neutral.

Properties of Minerals Observed with Crossed Nicols

(a) *Polarization Colours*. The colours seen in thin sections when viewed between crossed nicols are termed *polarization colours* and, apart from being the most immediately striking of the phenomena revealed by the polarizing microscope, are also amongst the most useful for mineral identification.

As we have seen, light from an original single, plane-polarized beam passes through anisotropic crystals as two rays with different velocities. When the nicols are crossed the two rays are recombined into a single plane by the analyser and are in a position to interfere with one another, since they will now in general be out of phase. The phase difference depends on three factors:

(i) The thickness of the section.
(ii) The wavelength of the light.
(iii) The difference in refractive index between the two rays. In most minerals this is substantially constant for all wavelengths.

Thus in a given grain, wavelength is the only variable and the phase difference will vary for the different wavelengths in the white light supplied. Hence certain colours will be reinforced because of path differences which happen to coincide with a whole number of wavelengths, and others will be cut out because of path differences involving a half wavelength. The result is a coloured ray, i.e., a coloured appearance of the grain, due to the removal of certain wavelengths from the original white.

Consider now the effect of progressively increasing the thickness of a given grain, an effect which can be observed by examining a section cut into a wedge. It is noticed that the colours change with increasing thickness in a repeating pattern which is known as Newton's Scale (see Fig. 6.7). The scale is divided into orders, and with the exception of the first order in which the grey and grey-white colours are unique, each consists of a set of colours somewhat akin to those of the ordinary white-light spectrum. The second-order colours are pure and bright and approach the spectral colours most closely, while those of the third order have about them what might be termed a distinct cosmetic tinge. The higher-order colours are paler and consist of alternating bands of pink and green which eventually become quite indistinguishable and form a pearly pale grey termed high-order white.

The effect of varying the thickness is exactly the same as varying the difference between the refractive indices of the two rays while keeping the thickness constant. In a given mineral this effect is seen by observing sections which have different orientations. For example, if a mineral such as augite, which is easily identifiable in plane polarized light, is examined between crossed nicols it will be seen that different grains show different polarization colours. Assuming that the thin section has been cut to the standard thickness of 0·03 mm, careful study of a large number of grains will show that all the colours of Newton's Scale up to the middle of the second order are present, but that there are no colours above this. Conversely, certain grains can be found which show almost no transmission of light, i.e., they show the black at the bottom of the first order. These sections are effectively isotropic and are cut across a direction known as the *optic axis*. Tetragonal, trigonal, and hexagonal minerals have one optic axis which lies parallel to the *c*-crystallographic axis. Sections of these minerals show their highest polarization colours in sections cut parallel to the *c*-axis. Monoclinic, orthorhombic, and triclinic minerals have two optic axes, neither of which coincides with a crystallographic axis, and show their highest polarization colours when both optic axes lie in the plane of the section.

To summarize, any mineral occurring in randomly oriented grains in a section of constant thickness will exhibit all the colours of the scale up to but not exceeding a certain maximum. It is this maximum colour which is likely to be a diagnostic character of the mineral and it is found in sections which show simultaneously the mineral's maximum and minimum refractive indices. The numerical difference between the two indices is spoken of as the *birefringence* of the mineral, e.g., quartz with a minimum index of 1·544 and a maximum of 1·553 has a birefringence of 0·009 which corresponds with a polarization colour of first-order white in sections having the standard thickness of 0·03 mm. Similarly, muscovite with indices of 1·56 and 1·60 has a birefringence of 0·04 and shows third-order blue as its maximum colour.

(b) *Determination of Birefringence*. The numerical value of the birefringence may easily be determined if the maximum polarization colour is identified by examining a number of differently oriented grains. First-order and second-order colours may with a little practice be identified at sight, but with colours of higher orders it is best to use the method of fringe-counting. It is necessary for this method to find a grain which is slightly cracked or is at the edge of the slide. Where the grain is in contact with the balsam the edge is very frequently crudely wedge-shaped in cross-section and careful study, if necessary using a high-power objective, will show a series of fringes, corresponding with all the colours in the scale *below* the colour shown by the main body of the crystal. The different orders are marked off by dark fringes which, on close examination, can be seen to be the red-violet-blue band forming the top of one order and the bottom of the next. It is then simply necessary to count the number of dark fringes on the thin edge of the crystal in order to determine the order of the main colour.

The thickness can be checked if necessary by determining the maximum colour of a known mineral in the slide, and the birefringence of the unknown mineral can then be determined from the Newton's Scale given in Fig. 6.7. The graphical procedure used is shown in Fig. 6.8.

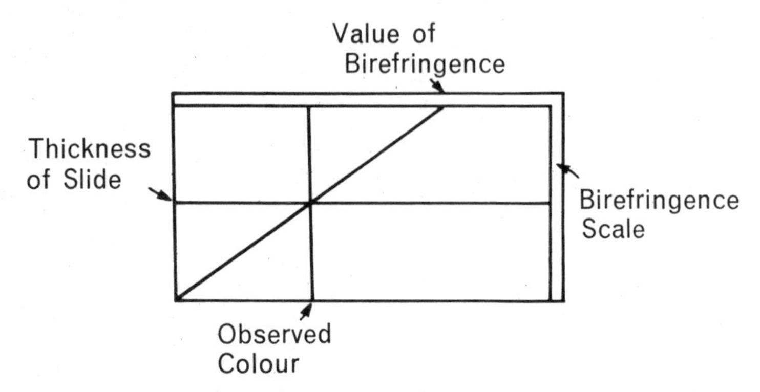

6.8 Principle of using the colour chart (Fig. 6.7) in the determination of birefringence or thickness.

(c) *Extinction Angles.* The extinction angle of a given grain is the angle between any specified crystallographic direction and either of the two vibration directions. This can be a very important distinguishing character for different minerals.

To take an example, nearly all micas (e.g., muscovite, biotite, chlorite) have what is termed straight extinction on the cleavage. Sections of these minerals cut so that the basal cleavage is visible, are always found to be in an extinction position when the cleavage trace (i.e., the line formed by the intersection of the cleavage and the slide) is parallel to the cross-wire. This is an example of an extinction angle of zero degrees.

In contrast, hornblende sections showing one cleavage trace or an elongated form (sections cut approximately parallel to the *c*-axis) usually have what is termed oblique extinction, that is to say that when the cleavage trace or length of the crystal is parallel to the cross-wire the grain is not in extinction. It is characteristically found with hornblende that a further rotation of the stage through 10–20 degrees is necessary to reach the extinction position. While this obliquity of extinction readily distinguishes hornblende from biotite, the size of the extinction angle also serves to distinguish it from pyroxenes, since the latter when showing one cleavage trace usually have extinction angles of the order of 45 degrees. A comparison of biotite, hornblende, and augite sections is given in Fig. 6.9.

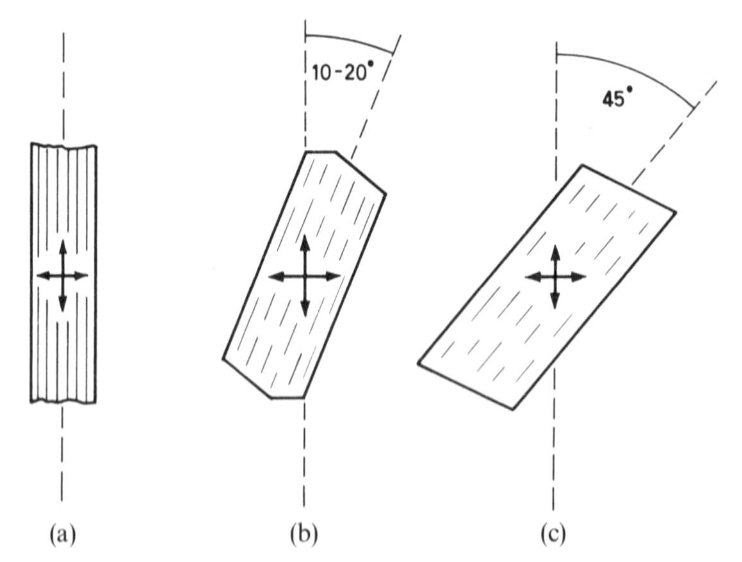

6.9 Maximum extinction angles of elongated sections of (a) biotite, (b) hornblende, (c) augite. Vibration directions shown by arrows. Note that in the correct orientation to show the maximum extinction angle, hornblende and augite show only slight traces of cleavage because the cleavage planes are not normal to the slide.

The positions of vibration directions relative to crystallographic directions can also, however, be of great use in determining the crystal system of an unknown mineral since in a general way the crystallographic symmetry is matched by an optical symmetry. Crystals belonging to the tetragonal, trigonal, and hexagonal systems have a large number of crystallographic symmetry elements and therefore show highly symmetrical optical properties. We have already seen that sections cut normal to the *c*-axis always have nil birefringence. Any other section always shows straight extinction relative to the trace of the *c*-axis. Thus any crystal belonging to these systems will often show straight extinction relative to the crystal outline provided that prism or pinacoidal faces are prominent. Extinction will also similarly be parallel to the trace of prismatic or pinacoidal cleavages. Tourmaline and apatite, and the micas (which are pseudo-hexagonal) all provide good examples of this type of extinction.

Orthorhombic crystals, while of lower symmetry than those mentioned, nevertheless show straight or symmetrical extinction in many sections. Extinction is strictly of this type whenever a symmetry plane is normal to the plane of the slide, and in other orientations extinction angles usually remain small. Olivine is a commonly encountered mineral which shows this behaviour.

Monoclinic crystals such as clinopyroxenes or hornblendes possess a single symmetry plane parallel to (010). Hence only those sections which are cut normal to this direction show straight or symmetrical extinction. Sections cut parallel to the symmetry plane, in contrast, show a maximum value for the extinction angle relative to the cleavage trace marking the trace of the *c*-axis. Other sections in which the trace of *c* is still observable, i.e., sections cut approximately parallel to the axis with the symmetry plane lying oblique to the plane of the slide, show an extinction angle which is less than the maximum. Thus, the extinction angle shows a behaviour analogous to that of the polarization colours, and in the determination of the angle for a given mineral, a number of grains must be studied in order to find the maximum. It is this maximum value which is meant when we state that a mineral (as opposed to a particular grain) has an extinction angle of so-many degrees.

Triclinic crystals, as one might expect, show no regularity of extinction, but the extinction angles of one group, the plagioclase feldspars, are of extreme importance.

The plagioclase feldspars form complete solid-solutions between the two end-members, albite ($NaAlSi_3O_8$) and anorthite ($CaAl_2Si_2O_8$). The composition of a particular plagioclase is expressed by the molecular percentage of anorthite it contains, e.g., the notation An_{100} refers to pure anorthite, An_{60} refers to a feldspar which consists of 60 per cent anorthite and 40 per cent albite. Since plagioclase feldspars are such prominent and common constituents of rocks it is important to determine whenever possible what sort of plagioclase is present. The microscopist who cannot distinguish albite from anorthite may obviously gain a very false impression of the composition of a rock he examines.

Fortunately extinction angles provide a ready means of determining plagioclase compositions with fair accuracy, but since the extinction angle of a grain is dependent on the orientation it is necessary to have a standard orientation to which all plagioclases are referred. Thus, with the method of determination we shall employ it is necessary to identify grains which lie with the (010) plane normal to the plane of the slide. These sections are, however, easily identifiable because they show the twin lamellae of albite-law twins sharply and clearly (see Figs. 7.11 and 7.12). Although albite-law twins are by far the most common lamellar twins observed in plagioclase it is necessary to carry out the following simple test to avoid possible confusion with other twin laws. Set the trace of the twins parallel to a cross-wire between crossed nicols and then rotate the specimen through 45 degrees. If the twins under observation are of the albite type the twinning will effectively disappear in this position because the two sets of lamellae show equal illumination.

When a suitable section has been located, the extinction angle is measured as the angle between the trace of the lamellae and, for our purposes, the nearest vibration direction. Ideally, when the section is cut strictly normal to (010), the two sets of lamellae will give identical extinction angles in opposite senses (see Fig. 6.10). In practice both angles are measured and if they are within 5 degrees of each other the average value is taken.

It will be found, however, that grains of plagioclase all having the

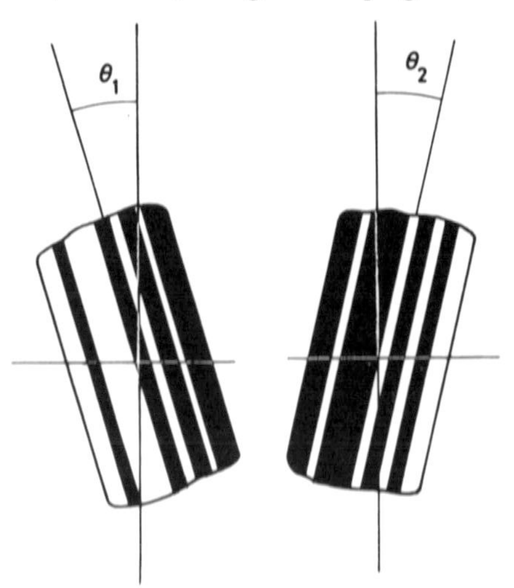

6.10 Determination of extinction angles in plagioclases cut normal to the plane of albite twin lamellae (010). The value taken is the average of θ_1 and θ_2.

same compositions will give differing extinction angles because their orientations vary. The grains selected should all have the (010) plane normal to the slide but have all possible orientations within this limitation. There is, however, one particular orientation which gives a *maximum* extinction angle, and it is this maximum value which is diagnostic of the composition of the feldspar. On the ordinary petrological microscope therefore, it is necessary to examine a large number of grains within one slide in order to be reasonably certain of finding this maximum value. This is of course directly analogous to the method of determining birefringence by finding the highest polarization colour shown.

It will be appreciated that this method does not allow the determination of extinction angles of more than 45 degrees. A more rigorous method measures the angle between the trace of the cleavage and the vibration direction corresponding with the ray of least refractive index. Plagioclases more calcic than An_{80} have extinction angles of more than 45 degrees on this basis and the simplified method recommended here will therefore confuse them with less calcic feldspars, e.g., a plagioclase with an extinction angle of 50 degrees will appear to have an extinction angle of only 40 degrees because our method measures not to the ray with least refractive index but to the *nearest* ray. Fortunately plagioclases of this composition are sufficiently rare for this to be a minor objection to the simplified method.

In Fig. 7.14 (p. 122) a curve relating the maximum extinction angle in sections cut perpendicular to (010) to the composition of the plagioclase is given. It will be observed that the curve gives an ambiguous result for extinction angles of less than 20 degrees, and that a particular composition, about An_{20}, has straight extinction for sections cut normal to (010). The ambiguity is fortunately readily resolved in most cases because plagioclases more sodic than An_{13} have refractive indices entirely below balsam. Thus a plagioclase with extinction angle of 10 degrees has the composition An_{12} if its refractive indices are below balsam, and the composition An_{28} if the refractive indices are above balsam.

7. Descriptions of minerals

Use of the Tables

Because of the variability of mineral characters in hand-specimens no single determinative table or scheme will suffice for mineral identification. Six tables for hand-specimen characters have therefore been included and, depending on the characters shown by an individual specimen, their use should enable the identity of an unknown mineral to be reduced to a relatively small number of possibilities.

In thin-section work it is, however, often possible to proceed according to a definite scheme of identification such as that which appears on pp. 99 to 100.

The mineral descriptions are arranged on a chemical basis as follows:

Native Elements
Sulphides
Halides
Oxides
Carbonates
Sulphates
Phosphates
Tungstates
Silicates

For minerals commonly encountered in rocks a description of both hand-specimen and thin-section characters is given. For other minerals only hand-specimen descriptions are included. Abbreviations used are H.–hardness, and S.G.–specific gravity.

Table 7.1

MINERALS COMMONLY FOUND WELL CRYSTALLIZED

Cubic System	*Trigonal System*
Galena	Hematite
Sphalerite	Calcite
Pyrite	Dolomite
Cuprite	Siderite
Magnetite	Tourmaline
Chromite	Quartz
Halite	(hexagonal appearance)
Fluorite	
Leucite	*Orthorhombic System*
Sodalite	Sulphur
Garnet	Anhydrite
	Barite
Tetragonal System	Aragonite
Rutile	Staurolite
Cassiterite	Topaz
Idocrase	
Zircon	*Monoclinic System*
	Gypsum
Hexagonal System	Augite
Apatite	Hornblende
Beryl	Orthoclase
Nepheline	

Only those minerals are included which often show a sufficiently good crystal form or cleavage to make the determination of crystal system in hand-specimens feasible for the student.

Table 7.2

MINERALS ARRANGED ACCORDING TO HARDNESS

Mohs Scale	*Minerals of Approximately Equal Hardness*
1. TALC	Graphite, clay minerals, chlorite, pyrolusite.
2. GYPSUM	Sulphur, halite, muscovite, chlorite, stibnite, galena, cinnabar, realgar, orpiment.
3. CALCITE	Anhydrite, barite, biotite, copper, silver, gold, bornite.
4. FLUORITE	Siderite, dolomite, rhodocrosite, aragonite, sphalerite, cuprite, malachite, chalcopyrite, pyrrhotite.
5. APATITE	Serpentine, limonite (both can be softer), chromite, scheelite, wolframite, kyanite (parallel to elongation).
6. FELDSPARS	Nepheline, leucite, sodalite, pyroxenes, amphiboles, epidote, rutile, idocrase, arsenopyrite, pyrite, hematite (can be lower).
7. QUARTZ	Kyanite (across the elongation), andalusite (lower when weathered), garnet, olivine, staurolite, cassiterite.
8. TOPAZ	Beryl.
9. CORUNDUM	
10. DIAMOND	

Table 7.3

MINERALS WITH PROMINENT CLEAVAGES

(a) *in one direction only*
- topaz
- micas (muscovite, biotite, chlorite, etc.)
- orpiment
- stibnite
- wolframite
- gypsum

(b) *in two directions*
- kyanite
- hornblende (124°)
- pyroxenes (90°)
- feldspars (at or near 90°)

(c) *in three directions*
- (i) mutually perpendicular
  - halite galena anhydrite
- (ii) not mutually perpendicular
  - rhombohedral carbonates (calcite, dolomite, etc.)
  - barite (2 at 78°), the third perpendicular to these two

(d) *in more than three directions*
- fluorite—four directions, octahedral
- sphalerite—six directions, dodecahedral

Table 7.4

HEAVY MINERALS WITH APPROXIMATE SPECIFIC GRAVITIES

Mineral	S.G.	Mineral	S.G.
Corundum	4·0	Bornite	5·1
Chalcopyrite	4·2	Magnetite	5·2
Rutile	4·2	Cuprite	6·0
Barite	4·5	Cassiterite	7·0
Stibnite	4·6	Wolframite	7·2
Chromite	4·6	Galena	7·5
Ilmenite	4·7	Cinnabar	8·1
Pyrolusite	4·7	Copper	8·9
Pyrite	5·0	Silver	10·5
Hematite	5·0	Gold	19·3

Table 7.5

MINERALS WITH COLOURED STREAKS

Black or Grey
Pyrolusite, psilomelane, graphite, galena, bornite, chalcopyrite, pyrite, stibnite, arsenopyrite, magnetite, ilmenite.

Brown, Dark Brown
Wolframite.

Light Brown
Sphalerite, cassiterite, rutile.

Red
Cuprite, cinnabar, hematite.

Other Colours
Limonite (yellow ochre), sulphur, orpiment (yellow), malachite (light green), azurite (light blue), copper (copper coloured), silver (silver), gold (golden).

Table 7.6

MINERALS HAVING RELATIVELY CONSTANT AND DISTINCTIVE COLOURS

Gold	Cinnabar (red)
Silver	Cuprite (red)
Copper	Realgar (orange)
Bornite (iridescent tarnish)	Orpiment (yellow)
Chalcopyrite (iridescent tarnish)	Sulphur (yellow)
Sphalerite (whitish metallic)	Malachite (bluish green)
Marcasite (pale brass yellow)	Epidote (yellowish green)
Pyrite (brass yellow)	Azurite (blue)
Chalcopyrite (deep brass yellow)	Kyanite (mottled blue)

Table 7.7

SCHEME OF MINERAL IDENTIFICATION IN THIN SECTIONS

Use of the table. The system depends on the estimation first of the birefringence of the unknown mineral (see p. 89), followed by a determination of the relief. Such features as cleavage, twinning, and pleochroism are then used to reduce the possibilities to a manageable number. These may then be looked up in more detail in the section on mineral descriptions which follows.

Birefringence	*Relief*	*Other Features*	*Mineral*
Nil Polarization colours—black	Moderate to low negative		Analcite
	High positive		Garnet
Very Low below 0·001 Polarization colours up to 1st-order dark grey	Moderate to low negative	No twinning	Analcite
		Lamellar twinning in sectors	Leucite
	Low positive	Micaceous form	Chlorite
Low 0·001–0·010 Polarization colours up to 1st-order pale yellow	Low negative	No lamellar or cross-hatched twinning	Nepheline Orthoclase
		Cross-hatched twinning	Microcline
		Lamellar twinning mainly in one direction	Albite (Plagioclase)
	Low positive	No cleavage or twinning	Quartz
		Lamellar twinning mainly in one direction	Oligolcase-Anorthite (Calcium bearing plagioclase)
	High positive		Orthorhombic Pyroxenes Andalusite

Table 7.7 (*contd.*)

Birefringence	*Relief*	*Other Features*		*Mineral*
MODERATE 0·010–0·030 Polarization colours up to mid second order	High positive	Marked cleavages	Pleochroic	Most AMPHIBOLES AEGIRINE (Clino-pyroxene)
			Non-pleochroic (usually colourless)	Most CLINOPYROXENES TREMOLITE (Amphibole) KYANITE
		Cleavages not marked	Pale yellow pleochroism	STAUROLITE
			Fibrous or acicular	SILLIMANITE
HIGH 0·030–0·040 Polarization colours up to mid 3rd order	High positive	One very marked cleavage	Pleochroic	BIOTITE
			Colourless	MUSCOVITE
		Cleavage not prominent		OLIVINE EPIDOTE
VERY HIGH above 0·040	Moderate positive			CALCITE DOLOMITE
Polarization colours up to 'high-order white'	Very high positive			ZIRCON SPHENE

Native Elements

NATIVE GOLD Au　　Cubic

Crystallography. Crystals rare. Usually in shapeless grains and nuggets and in filiform masses.

Colour—golden. *Streak*—golden. *Lustre*—metallic, opaque. *H.*—2·5–3. *S.G.*—20 or less, depending on the amount of impurity. *Fracture*—hackly. Also highly malleable and sectile.

Distinguishing features. Cannot easily be confused with other minerals such as pyrite and chalcopyrite because of its softness and sectility. The colour too is distinctive, since gold is much more golden than any of the brassy coloured sulphides.

Occurrence. As flakes in river gravels, etc. (placer deposits). Also associated with quartz as a vein mineral.

NATIVE SILVER Ag — Cubic

Crystallography. Usually filiform or massive.

Colour—silver. *Streak*—silver. *Lustre*—metallic, opaque. *H.*—2·5–3. *S.G.*—10–11. *Fracture*—hackly. Also highly malleable and sectile.

Distinguishing features. Distinguished from native gold by the colour.

Occurrence. Mostly in veins associated with silver sulphides, cobalt and nickel minerals, or native copper.

NATIVE COPPER Cu — Cubic

Crystallography. Crystals rare. Usually massive or in ramifying arborescent masses.

Colour—coppery. *Streak*—coppery. *Lustre*—metallic, shining to dull. *H.*—2·5–3. *S.G.*—9. *Fracture*—hackly. Also malleable.

Distinguishing features. Distinguished from gold and silver by colour.

Occurrence. In vein deposits associated with other copper minerals, and in the amygdales of certain basalts.

GRAPHITE C — Hexagonal

Crystallography. Crystals rare. Usually in scaly masses.

Cleavage. Perfect parallel to the scales. Also sectile.

Colour—dark grey to black. *Streak*—black, shiny. Soft enough to mark paper. *Lustre*—metallic. *H.*—1·5. *S.G.*—2.

Distinguishing features. The form distinguishes it from stibnite, which is also somewhat harder. Graphite is easily confused with the molybdenum sulphide, molybdenite.

Occurrence. Mainly disseminated in metamorphic rocks.

SULPHUR S — Orthorhombic

Crystallography. Well-formed steep bipyramidal crystals occasionally found (see Fig. 7.1). Often in encrustations in which individual crystals are not visible.

Cleavage. Poor parallel to {110} and {111}.

Colour—yellow. *Streak*—yellow. *Lustre*—resinous, transparent to translucent. *H.*—1·5–2·5. *S.G.*—2.

Distinguishing features. Could be confused with orpiment but the latter has a higher S.G. Sulphur can easily be melted and burns with a characteristic smell.

Occurrence. Mainly as encrustations in volcanic areas. Also interbedded with sediments and evaporites in areas of salt domes.

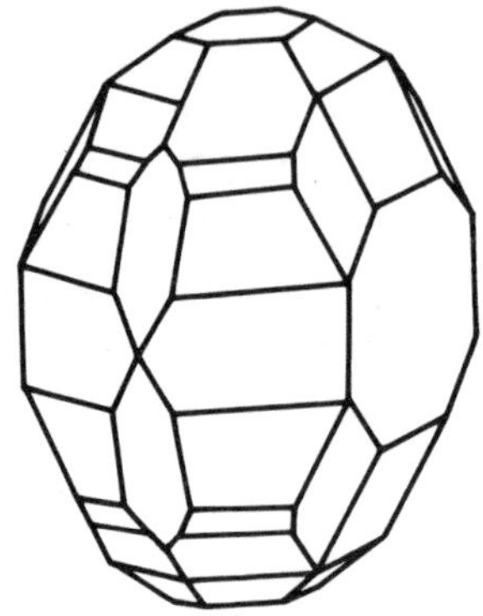

7.1 A crystal of sulphur showing a great variety of forms, mainly pyramidal.

Sulphides

BORNITE Cu_5FeS_4 Cubic

Crystallography. Crystals rare. Usually massive.

Colour—bronze when fresh but tarnishes rapidly to a purplish, iridescent colour. The tarnished surface is very characteristic and has given rise to the term 'peacock ore'. *Streak*—greyish black. *Lustre*—metallic. *H.*—3. *S.G.*—about 5·1.

Distinguishing features. The purple tarnish is characteristic but could cause confusion with chalcopyrite which also shows this feature. The latter, however, has a much more golden-yellow colour on fresh surfaces.

Occurrence. An important ore of copper, mainly found as a vein mineral.

CHALCOPYRITE $CuFeS_2$ Tetragonal

Crystallography. Usually massive. Sometimes as crystals similar to tetrahedron.

Colour—brass yellow, often with a purplish iridescent tarnish. *Streak*—greenish black. *Lustre*—metallic. *H.*—3·5–4. *S.G.*—4·2.

Distinguishing features. Can easily be confused with pyrite which has a similar brassy colour. Chalcopyrite is, however, much softer than pyrite (hardness 3·5–4 as opposed to hardness 6–6·5). Pyrite does not tarnish. Bornite is most easily distinguished from chalcopyrite by its more coppery colour when fresh, though both minerals have iridescent tarnishes on weathered surfaces.

Occurrence. A vein mineral. The most important ore of copper.

GALENA PbS — Cubic

Crystallography. Crystals common, usually cubes, octahedrons or combinations of different cubic forms. Also commonly in granular masses.

Cleavage. Perfect parallel to the faces of the cube.

Colour—lead grey. *Streak*—grey. *Lustre*—metallic. Bright when fresh. Dull on weathered surfaces. *H.*—2·5. *S.G.*—7·5.

Distinguishing features. Unlikely to be confused with any other common mineral. The high specific gravity, softness, lead grey colour, and perfect cubic cleavage are very distinctive.

Occurrence. A vein mineral. The chief ore of lead.

SPHALERITE ZnS (Zinc Blende) — Cubic

Crystallography. Crystals common with a variety of cubic forms including commonly the tetrahedron.

Cleavage. Perfect dodecahedral (six directions).

Colour—very variable from pale yellow to reddish brown, dark brown, and black. *Streak*—white to reddish brown. *Lustre*—resinous. *H.*—3·5–4. *S.G.*—4·1.

Distinguishing features. Not always an easy mineral to identify because of its variable colour. Its association in veins with galena and pyrite together with its characteristic lustre will often be sufficient to identify it.

Occurrence. A vein mineral. The most important ore of zinc.

CINNABAR HgS — Trigonal

Crystallography. Usually massive or in encrustations.

Colour—bright red. *Streak*—characteristic vermilion colour. *Lustre*—adamantine to dull. *H.*—2·5. *S.G.*—8·1.

Distinguishing features. The colour may cause confusion with realgar (which has a much lower specific gravity) and cuprite (which is much harder than cinnabar).

Occurrence. The most important ore of mercury. Found in veins associated with other sulphides, and as impregnations.

PYRITE FeS_2 — Cubic

Crystallography. Commonly crystallized as striated cubes or pyritohedra (see Figs. 4.4 and 4.5). Also massive.

Colour—brass yellow. *Streak*—greenish black. *Lustre*—metallic. *H.*—6–6·5. *S.G.*—5·0.

Distinguishing features. The crystals when well formed are very distinctive. May be confused with gold and chalcopyrite, both of which are much softer.

Occurrence. A very common vein mineral associated with other sulphides. Also found as crystals in shales and schists and as a replacement of fossils. A relatively rare accessory constituent of igneous rocks.

Thin section characters. Not infrequently encountered in thin sections of rocks where it appears as an opaque mineral. It is necessary to raise the light source so that light is reflected off the upper surface of the slide in order to observe the distinctive brassy colour. This distinguishes it from magnetite and ilmenite which have a steely grey appearance under these conditions, but will not serve to distinguish pyrite from chalcopyrite.

MARCASITE FeS_2 — Orthorhombic

A mineral similar in many respects to pyrite from which it may be distinguished by its much paler, whitish, brassy colour. Found in radiating concretionary masses or curving cockscomb groups in sedimentary rocks. Used as a material for jewellery.

ARSENOPYRITE FeAsS — Monoclinic

Crystallography. Mainly massive.

Colour—silvery white. *Streak*—greyish black. *Lustre*—metallic. *H.*—5·5–6. *S.G.*—3·6.

Distinguishing features. When struck with a hammer gives off a noticeable garlic-like smell. Not easily confused on account of its colour with other common ore minerals, but is very similar in appearance to some of the rarer ones, such as arsenides of nickel.

Occurrence. A common vein mineral. An ore of arsenic.

REALGAR AsS Monoclinic

Crystallography. Crystals rare. Usually massive.

Colour—orange red. *Streak*—red to orange. *Lustre*—resinous, transparent to translucent. *H*—1·5–2. *S.G.*—3·6. Sectile.

Distinguishing features. Distinguished from cinnabar by its more orange colour, extreme softness and relatively low specific gravity.

Occurrence. As veins and impregnations in rocks, often associated with orpiment. An important ore of arsenic.

ORPIMENT As_2S_3 Monoclinic

Crystallography. Crystals rare, commonly massive and in mica-like masses.

Cleavage. Perfect parallel to (010). Cleavage flakes are sectile and flexible.

Colour—bright yellow on fresh surfaces. *Streak*—yellow. *Lustre*—pearly to resinous. *H*.—1·5–2. *S.G.*—3·5.

Distinguishing features. Only likely to be confused with native sulphur, but the mica-like cleavage of orpiment is distinctive and it is much denser than sulphur.

Occurrence. A vein mineral, associated with realgar.

STIBNITE Sb_2S_3 Orthorhombic

Crystallography. Commonly in elongated prismatic crystals with longitudinal striations. Often in radial groups. Also in columnar and bladed masses.

Cleavage. Perfect parallel to (010).

Colour—dark grey. *Streak*—grey. *Lustre*—metallic when fresh. *H*.—2. *S.G.*—4·6.

Distinguishing features. The form is distinctive, and the extreme softness distinguishes it from most other similar minerals, e.g., wolfram.

Occurrence. A vein mineral and ore of antimony.

Halides

HALITE $NaCl$ Cubic

Crystallography. Crystals common, usually cubes.

Cleavage. Perfect parallel to the cube faces.

Colour—often colourless or tinged with brown, yellow, or occasionally blue. *Streak*—colourless. *Lustre*—vitreous, transparent to translucent. *H.*—2, easily scratched with the finger nail. *S.G.*—2·2. *Taste*—salty when licked.

Distinguishing features. Extreme softness, cubic crystals, cleavage, and taste. An easy mineral to recognize.

Occurrence. In evaporite deposits. An important economic mineral.

FLUORITE CaF_2 Cubic

Crystallography. Crystals common, usually cubes. Also granular and massive.

Cleavage. Perfect parallel to {111} giving four cleavage directions (see Fig. 5.1).

Colour—extremely variable; blue, violet, green, yellow, pink, colourless. *Streak*—colourless. *Lustre*—vitreous, transparent to translucent. *H.*—4. *S.G.*—3·2.

Distinguishing features. Usually recognized by the crystal form, the way it is possible to cleave off the corners of the cube also being distinctive. But be careful not to ruin a good specimen.

Occurrence. A common vein mineral, often associated with galena and sphalerite. An important economic mineral with applications as a flux in steel making, for optical purposes, and, for certain deeply coloured varieties, as a semi-precious stone used for carving.

Oxides

QUARTZ SiO_2 Trigonal

Crystallography. Crystals common, prismatic, with pseudo-pyramidal terminations (see Fig. 7.2). For a fuller discussion see p. 63.

Cleavages. None.

Colour—usually colourless or white. Coloured varieties include amethyst (purple), rose quartz (pink), cairngorm (yellow-brown), and smokey quartz (dark to grey). These are semi-precious in good transparent specimens. *Fracture*—conchoidal. *Streak*—white. *Lustre*—vitreous. *H.*—7. *S.G.*—2·6.

Optical properties. Colour—colourless in thin section. *Form*—usually in anhedral grains. Occasionally, as a phenocryst in volcanic rocks, may show an equidimensional euhedral form (bi-pyramidal). *Cleavage*—none. *Relief*—low positive. *Birefringence*—low (0·009). Maximum interference colour is first-order white faintly tinged with yellow. Most sections show greys and white. *Extinction*—parallel or symmetrical with respect to crystal edges. Not normally determinable because of the anhedral form.

Distinguishing features. Easy to identify in hand-specimens by the form of the crystals. When massive its superior hardness distinguishes it from many other white minerals. In hand-specimens of rocks it is usually more transparent and glassy-looking than feldspar while nepheline, which may resemble it when colourless, has a distinctive greasy lustre. In fractured specimens the conchoidal fracture is distinctive.

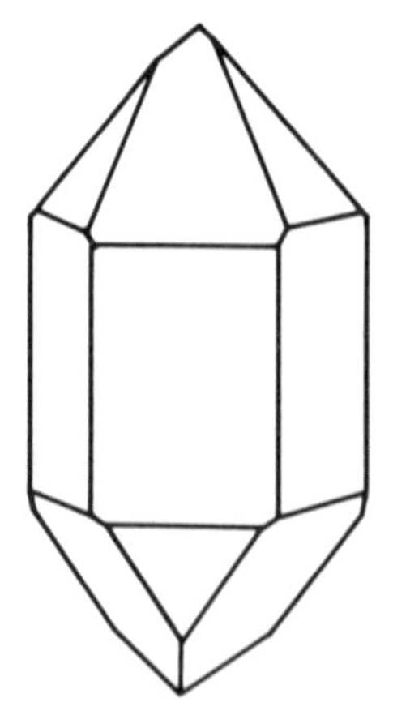

7.2 Quartz.

In thin section, quartz may be difficult to identify because of its lack of positive features.* It is distinguished from feldspars by its clarity and freedom from alteration and has a notably different refractive index from alkali feldspars (quartz has low positive relief while alkali feldspars have low negative relief). Quartz also has slightly higher birefringence than most feldspars, which in general give grey, rather than white, first-order colours. Apatite resembles quartz but has a much higher relief. Nepheline resembles quartz very closely but has a slightly lower birefringence and often shows a distinct cleavage.

Occurrence. One of the most common of all minerals, quartz is the principal constituent of sandstones and is present in many other sedimentary rocks as detrital grains and as a cementing material. Quartz is also abundant in many metamorphic rocks and is an essential constituent of the granitic igneous rocks. In addition to these, quartz is probably the most common of the vein minerals.

* The identification of quartz is best checked by means of interference figures.

CORUNDUM Al_2O_3 Trigonal

Crystallography. Six-sided crystals, sometimes elongated into pyramidal forms (see Fig. 7.3). Also in six-sided plates.

Colour—colourless, grey, bluish, violet, reddish. *Streak*—colourless. *Lustre*—vitreous. *H.*—9. *S.G.*—4·0.

Distinguishing features. Resembles some silicates but the hardness is distinctive as is the high specific gravity.

Occurrence. In pegmatites and some aluminous schists. Rarely as an accessory in igneous rocks such as nepheline syenite. Varieties of corundum such as ruby and sapphire are valuable gem stones. Emery, used as a polishing compound for hard materials, is an impure form of corundum.

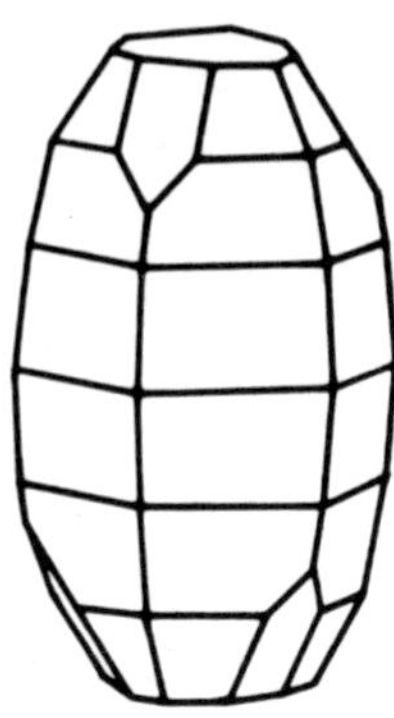

7.3 A crystal of corundum showing a number of steep pyramidal forms.

CUPRITE Cu_2O Cubic

Crystallography. Crystals commonly cubes, octahedrons, rhombdodecahedrons, and combinations of these forms. Also massive.

Colour—shades of red, often bright red. *Streak*—brownish red. *Lustre*—adamantine, sub-metallic, dull. *H.*—3·5–4. *S.G.*—6·0.

Distinguishing features. Usually associated with more easily recognizable copper minerals which should prevent confusion with cinnabar.

Occurrence. A widespread copper ore associated in veins with malachite, azurite, and native copper.

MAGNETITE Fe_3O_4 Cubic

Crystallography. Crystals not uncommon, usually octahedra. Also massive and granular.

Colour—black. *Streak*—black. *Lustre*—metallic to sub-metallic. *H.*—6. *S.G.*—5·2. Magnetic.

Distinguishing features. In specimens which are large enough a test with a magnet is sufficient to identify magnetite. Otherwise it is difficult to distinguish from spinel, and in the absence of octahedral crystals, from ilmenite.

Thin section characters. Opaque in thin section, it is necessary to raise the light source and view the specimen by reflected light. The appearance when viewed in this way is steely grey, in contrast to most of the sulphides, which show a more brassy colour. Ilmenite cannot however, always be separated from magnetite by this test.

Magnetite crystals are well formed in many rocks and give square or hexagonal outlines because of the octahedral form. Spinels are similar in appearance but may be translucent rather than opaque.

Occurrence. Magnetite is an extremely widespread though rarely abundant constituent of all classes of rocks. It may be concentrated in large masses as segregations in igneous bodies and as vein-like off-shoots. Magnetite is an important ore of iron.

HEMATITE Fe_2O_3 Trigonal

Crystallography. Crystals common, tabular rhombohedral forms (see Fig. 7.4) with scalenohedral faces forming thin lenticular scales, flat scales, etc. Also commonly in reniform or mammilary masses.

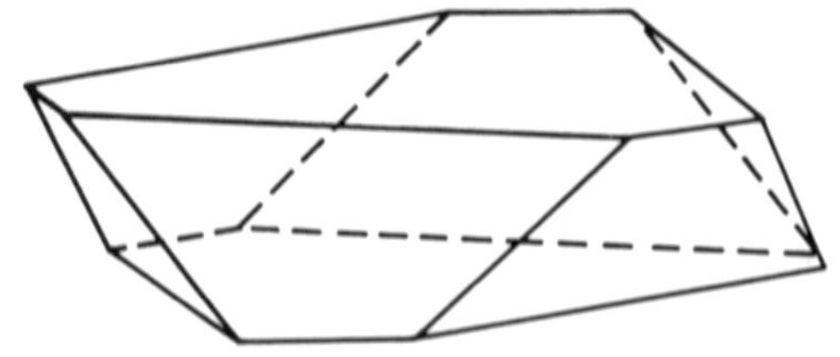

7.4 Hematite showing typical form of combined rhombohedron and basal pinacoid.

Colour—when well crystallized usually black. Reniform masses are dark red to reddish black. *Streak*—dark red. *Lustre*—black, well-crystallized specimens have splendent metallic lustres and this variety is sometimes known as 'specularite'. The reniform reddish varieties have a duller metallic to sub-metallic lustre. *H.*—very variable depending on the state of aggregation. Crystals have hardness 6, but aggregates may be as low as hardness 1. *S.G.*—about 5.

Distinguishing features. Usually an easy mineral to identify because of the distinctive form of the crystals and the appearance of the reniform masses. The red streak is the most important property in case of doubt, e.g., confusion with reniform limonite, or confusion of specular hematite with other black minerals.

Occurrence. The most important of the iron ores, very widespread particularly in sedimentary rocks, where it may form large discrete masses. It is an important rock-forming mineral in some schists and gneisses and an important mineral in residual lateritic deposits formed by the leaching of rocks under tropical weathering conditions.

LIMONITE Hydrated iron oxide Amorphous

Form. Commonly earthy or powdery. Also in stalactitic, botryoidal, or mammilated masses.

Colour—brownish black to yellow ochre. *Streak*—yellow ochre. *Lustre*—sub-metallic, silky occasionally, often dull. Opaque. *H.*—5 but may be much softer in aggregates. *S.G.*—about 3 to 4.

Distinguishing features. The habit may cause confusion with hematite or the manganese oxides but the streak excludes these (red for hematite, black for the manganese oxides).

Occurrence. Very widespread as an alteration product of other iron minerals. An important constituent of some sedimentary iron ores. Ochreous forms of limonite are used as pigments.

ILMENITE $FeO.TiO_2$ Trigonal

Crystallography. Tabular to scaly crystals. Also massive and granular.

Colour—black. *Streak*—black. *Lustre*—metallic to sub-metallic, opaque. *H.*—5–6. *S.G.*—about 4·7.

Optical properties. An opaque mineral commonly encountered in thin sections of rocks. If the light source is raised so that the mineral is viewed in reflected light ilmenite has a steely grey appearance similar to that of magnetite. It may often be distinguished from magnetite by its tendency to form skeletal, platy crystals (see Fig. 7.5) and by its association with the white opaque alteration product known as leucoxene.

Distinguishing features. Similar in hand-specimens to magnetite, except when well crystallized. Ilmenite is only weakly magnetic compared with magnetite.

Occurrence. A common accessory mineral in basic igneous rocks. Occasionally found as an important constituent of beach sands.

7.5 Ilmenite in thin section, showing opacity and skeletal form. × 30. Plane polarized light.

SPINEL $MgO . Al_2O_3$ with Cr and Fe bearing varieties Cubic

Crystallography. Usually in small octahedra.

Colour—black, dark green, reddish, bluish, or brownish. *Lustre*—glassy, transparent to opaque. *H.*—7·5–8. *S.G.*—3·5–4.

Optical properties. Similar to magnetite when opaque. Translucent varieties are easily confused with garnet but the latter is usually considerably more transparent and shows a more rounded form in thin section because of its rhombdodecahedral habit (cf. the octahedral habit of spinel).

Occurrence. A rare mineral found in marbles and in basic and ultrabasic igneous rocks. The Cr-rich variety, chromite, is a valuable ore of chromium and is found as layers in large gabbroic intrusions.

MANGANESE OXIDES (Pyrolusite, Psilomelane, Wad)

The manganese oxides, pyrolusite (MnO_2) and psilomelane and wad (hydrated manganese dioxide with varying amounts of barium and potassium oxides) are difficult to distinguish from each other and are treated collectively here.

Form. Crystals rare, usually massive, reniform, botryoidal or as radiating fibrous aggregates. Also in dendritic encrustations on joint planes in rocks.

Colour—black, grey. *Streak*—black. *Lustre*—metallic, sub-metallic, or dull. Opaque. *H.*—varies from about 6 to as little as 2, depending on the state of aggregation. *S.G.*—in the range 3 to 4·8.

Distinguishing features. The habit and streak are usually distinctive. The manganese oxides may be confused with dark varieties of limonite but the latter has a yellow ochre streak.

Occurrence. Widespread, formed by the alteration of other manganese minerals in oxidized mineral veins. Also common in sedimentary formations where manganese minerals are precipitated chemically and biochemically.

CASSITERITE SnO_2 — Tetragonal

Crystallography. Crystals consist of tetragonal prisms with pyramidal terminations (see Fig. 7.6). Geniculate twins common.

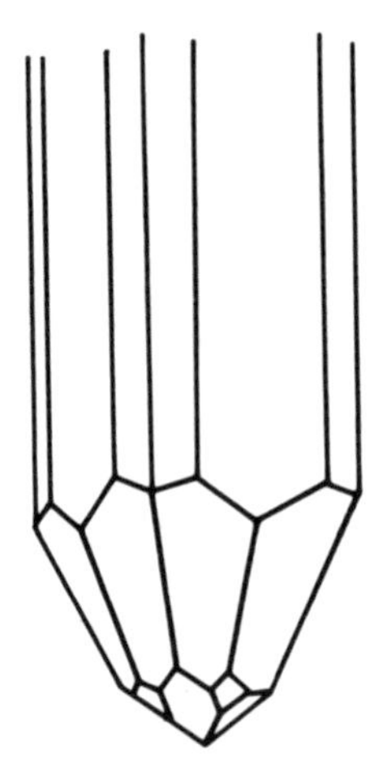

7.6 Cassiterite showing a number of tetragonal and ditetragonal prism and pyramid forms.

Colour—usually black or brown. *Streak*—white to pale brown. *Lustre*—adamantine, often brilliant. Translucent to opaque, depending on the darkness of the colour. *H.*—6–7. *S.G.*—7.

Distinguishing features. The hardness and high specific gravity distinguish cassiterite from most similar minerals.

Occurrence. Cassiterite is the most important ore of tin and is found in high-temperature veins and pegmatites, as well as in placer deposits.

Carbonates

MALACHITE $CuCO_3 . Cu(OH)_2$ — Monoclinic

Crystallography. Normally in massive, stalactitic, botryoidal or mammilary aggregates.

Colour—deep slightly bluish green. *Streak*—pale green. *Lustre*—somewhat silky in fibrous specimens. Usually dull. Mainly opaque. *H.*—3·5–4. *S.G.*—4·0.

Distinguishing features. The colour and streak coupled with the form of the aggregates are distinctive.

Occurrence. An important ore of copper, also used for ornamental purposes. Found disseminated through sediments and in the weathered zones of copper-bearing veins.

AZURITE $2CuCO_3 . Cu(OH)_2$ — Monoclinic

Crystallography. Equidimensional modified prismatic crystals. Commonly massive, earthy, or in encrustations.

Colour—deep azure. *Streak*—pale blue. *Lustre*—vitreous. Transparent to opaque. *H.*—3·5–4. *S.G.*—3·8.

Distinguishing features. The colour is very striking but may lead to confusion with less common blue copper minerals such as chalcanthite (copper sulphate) and linarite (sulphate of lead and copper).

Occurrence. In the weathered zones of copper-bearing veins, associated with malachite. A valuable ore of copper.

CALCITE $CaCO_3$ — Trigonal

Crystallography. Commonly well crystallized (see Figs. 4.14–4.16) having several habits including (a) combined scalenohedron $(21\bar{3}1)$ and prism $(10\bar{1}0)$, known as dog-tooth spar; (b) combined flat rhomb $(01\bar{1}2)$ and prism, known as nail-head spar.

Cleavage. Perfect parallel to the unit rhomb $(10\bar{1}1)$ (see Fig. 5.1).

Colour—usually colourless or white. *Streak*—white. *Lustre*—vitreous. *H.*—3. *S.G.*—2·7.

Optical properties. Colour—colourless in thin section. *Form*—usually in anhedral crystals. May be in oolites or spherulites or show organic structures, e.g., shell and coral fragments. *Cleavage*—easily visible in thin sections. Although there are three directions it is only possible to see two in any one section. These intersect at 75 degrees when normal to the plane of the slide. *Relief*—most grains show a strong change when rotated (twinkling), from high positive relief to low relief which may be positive or negative. *Birefringence*—extreme (0·172). Most grains show pearly-grey high-order colours. *Extinction*—symmetrical to the cleavage traces. *Twinning*—lamellar twinning common.

Distinguishing features. In hand-specimen the habit, and disposition of the rhombohedral cleavage make calcite relatively easy to recognize. When massive it is easily distinguished from quartz by its softness, and from barytes by its lower specific gravity. It is often, however, difficult to distinguish from other rhombohedral carbonates such as dolomite. In thin section the birefringence and twinkling are sufficient to distinguish calcite from other common minerals except dolomite.

Occurrence. Principal constituent of limestones and marbles. Common in igneous rocks as an amygdale mineral and as an alteration product. A very common vein mineral.

DOLOMITE $CaCO_3 . MgCO_3$. Trigonal

Very similar to calcite and difficult to distinguish from it. It is slightly harder than calcite (*H.*—3·5–4) and has a slightly higher specific gravity (*S.G.*—2·9). When well crystallized a rhombohedral form is common (contrast calcite) and often shows *curved* faces. The lustre is often somewhat more pearly than that of calcite.

SIDERITE $FeCO_3$ Trigonal

Similar to dolomite in habit and crystallography but often dark in colour (brownish and blackish). Distinguished from other rhombohedral carbonates by high specific gravity (*S.G.*—3·8).

RHODOCHROSITE $MnCO_3$ Trigonal

A rhombohedral carbonate mineral similar to calcite and dolomite. It may be distinguished from these by its greater specific gravity (about 3·5 in contrast to 2·7–2·8) and a characteristic rose-pink colour.

ARAGONITE $CaCO_3$ Orthorhombic

Crystallography. Crystals often prismatic with pointed domal terminations, frequently growing in radiating sheaves. Repeated twinning on the prism face gives crystals somewhat hexagonal in appearance (see Fig. 7.7). These are distinguished from true hexagonal crystals by the re-entrant angles between adjacent twins.

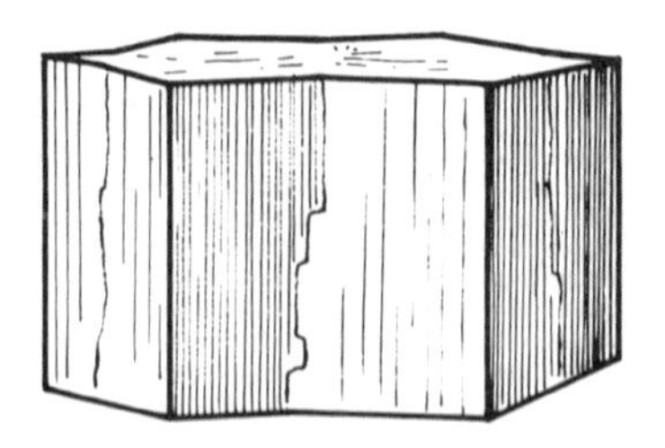

7.7 Repeated twin of aragonite. Note the slight re-entrant angles.

Cleavage. Poor parallel to (010).

Colour—white, grey, yellowish. *Streak*—colourless. *Lustre*—vitreous, transparent to translucent. *H.*—3·5–4. *S.G.*—2·9.

Distinguishing features. Best distinguished from calcite by the form of the crystals and the slightly superior hardness. In large crystals the poor quality of the cleavage and its single direction are also distinctive.

Occurrence. Within sedimentary rocks and as a hot-spring deposit. Also a prominent constituent of some shells, particularly corals.

Sulphates

GYPSUM $CaSO_4 . 2H_2O$ Monoclinic

Crystallography. Crystals common, combination of clinopinacoid, prism and hemipyramid. The crystals are frequently tabular parallel to (010) and may interpenetrate each other in an irregular fashion. Twins are not uncommon with either (100) or (101) as twin plane (see Fig. 4.22).

Cleavage. Perfect parallel to (010). It is possible to produce very thin flexible cleavage plates.

Colour—usually colourless or white. *Streak*—colourless. *Lustre*—vitreous to pearly. Transparent to opaque. *H.*—2, easily scratched with the finger nail. *S.G.*—2·3.

Varieties. Selenite is a term used for transparent, well-crystallized gypsum. Alabaster is a white, compact, massive variety. Satin Spar is a fibrous variety with a silky lustre.

Distinguishing features. Well-crystallized gypsum is easily identified. In massive varieties the extreme softness is the best guide.

Occurrence. In evaporite deposits and as crystals in clays and similar sediments.

BARITE $BaSO_4$ Orthorhombic

Crystallography. Crystals common, often tabular parallel to the basal pinacoid and bounded by prism and dome faces (see Fig. 7.8). Also commonly massive.

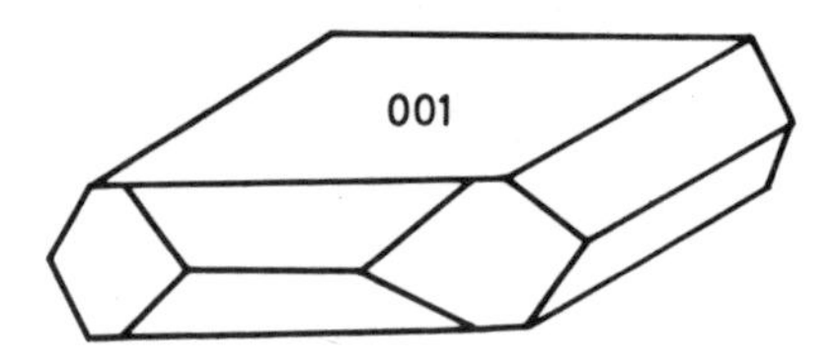

7.8 Tabular crystal of barite bounded by the basal pinacoid, domes and small prism faces.

Cleavage. Perfect parallel to the prism form {110} giving two cleavage directions at an angle of 78 degrees. Also perfect parallel to {001} giving a third direction normal to the other two (see Fig. 5.1).

Colour—colourless or white, pale pink, green, or bluish. *Streak*—colourless. *Lustre*—vitreous, pearly. Transparent to opaque. *H.*—3–3·5. *S.G.*—4·5.

Distinguishing features. Because of its pale colour and vitreous lustre barite may be mistaken for other similar minerals such as calcite, quartz, fluorite, and feldspar. The abnormally high specific gravity distinguishes it readily from these, and if additional tests are needed a careful examination of the cleavage will often reveal the orthorhombic symmetry (cf. fluorite, four cleavages on the octahedral form of the cubic system, calcite with three cleavage directions on a rhombohedral form). Feldspars and quartz are easily excluded by their superior hardness.

Occurrence. A common vein mineral, often associated with galena, sphalerite, fluorite, and quartz.

Phosphates

APATITE Calcium phosphate with fluorine and chlorine Hexagonal

Crystallography. Crystals common, prismatic, with pyramid and basal pinacoid forms.

Cleavage. Poor parallel to (0001).

Colour—variable. Pale green, bluish green, yellow, brownish, reddish. *Streak*—white. *Lustre*—vitreous. *H.*—5. *S.G.*—3·2.

Optical properties. Colour—colourless in thin section. Usually extremely clear. *Form*—often in small euhedral crystals showing elongated rectangular sections cut parallel to the *c*-axis, and perfect hexagonal sections cut normal to this direction (see Fig. 6.4). *Cleavage*—not usually noticeable in thin section. *Relief*—moderate positive. *Birefringence*—low (0·003). Polarization colours are greys of the first order. *Extinction*—parallel to elongation of crystal.

Distinguishing features. In hand-specimen distinguished from most other hexagonal minerals by its hardness. In thin section the form and clarity coupled with the moderate positive relief and weak birefringence make apatite an easy mineral to recognize.

Occurrence. A very common accessory mineral in igneous rocks. Also in veins and pegmatites.

Tungstates

WOLFRAMITE $(Fe,Mn)WO_4$ Monoclinic

Crystallography. Bladed crystals common, flattened parallel to (010).

Cleavage. Perfect parallel to (010).

Colour—black to reddish brown. *Streak*—chocolate brown. *Lustre*—metallic or sub-metallic on cleavage faces. Otherwise usually dull. *H.*—4·5–5. *S.G.*—6·0.

Distinguishing features. Usually relatively easy to identify because of the form of the crystals, the single prominent cleavage, and the brown streak.

Occurrence. A vein mineral. An important ore of tungsten.

Silicates

THE FELDSPARS

The feldspars are as a group the most important of the rock-forming minerals and consist of partly discontinuous solid solutions between the end members *orthoclase* ($KAlSi_3O_8$), *albite* ($NaAlSi_3O_8$), and *anorthite* ($CaAl_2Si_2O_8$). Figure 7.9 shows the relations between the

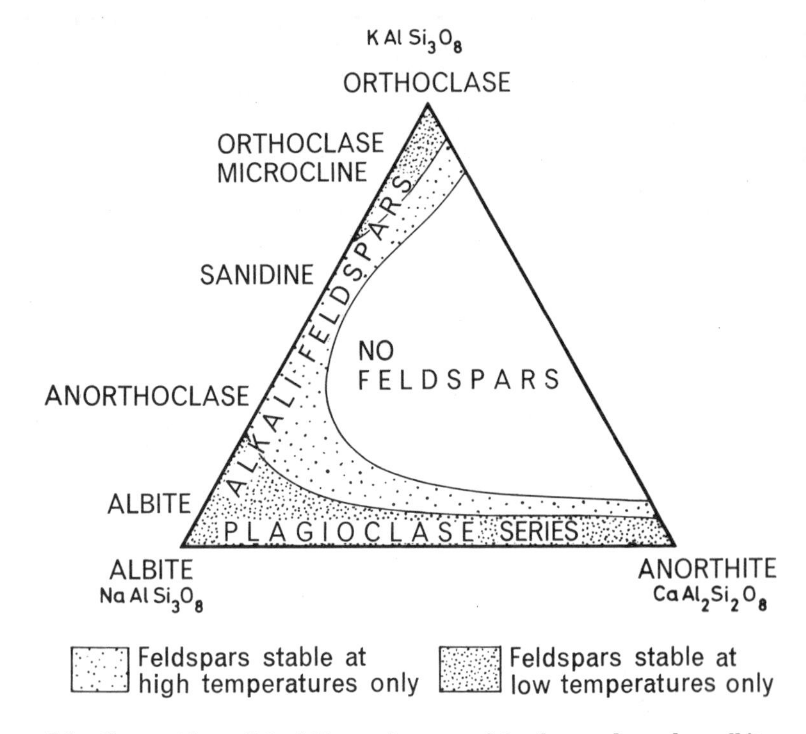

7.9 Compositions of the feldspars in terms of the three end members albite, anorthite, and orthoclase.

end-members in the form of a triangular diagram. Note that at all temperatures the extent of solid solutions between orthoclase and anorthite is small, while solution between orthoclase and albite is continuous only in feldspars stable at high temperatures, and is discontinuous in low-temperature feldspars. Albite and anorthite form complete solid solutions at all temperatures. The latter solution involves the substitution of CaAl in the lattice in place of NaSi whereas the solution of albite in the orthoclase lattice involves only the substitution of Na for K.

Hand-Specimen Characters

Crystallography. All the feldspars are very similar in hand-specimens and it is rarely possible to distinguish them on the basis of form alone. Orthoclase and sanidine are monoclinic and show {010}, {001}, {110} as common forms, together with hemi-domes such as $\{\bar{1}01\}$ and $\{\bar{2}01\}$ (see Fig. 4.20). The angle β in these feldspars is about 116 degrees. The triclinic feldspars, microcline and the plagioclases, are similar while showing an angle between (001) and (010) (90 degrees in the monoclinic feldspars) of about 86 degrees. All the feldspars show a perfect (001) cleavage and a slightly less perfect (010) cleavage.

Twinning. Twinning is very common in the feldspars and all varieties commonly exhibit simple twinning on the *Carlsbad* law, in which the composition plane is (010) and the twin plane (100) (see Figs. 4.23 and 7.10). Simple twins on the *Manebach* (composition plane and twin plane (001)), and *Baveno* laws (composition and twin plane

7.10 Sanidine crystal in trachyte, showing Carlsbad twinning. ×50. Crossed nicols.

(021)) are much less common. Plagioclases are in addition characterized by polysynethetic (repeated lammellar) twinning on (010) known as albite twinning (see Figs. 7.11 and 7.12). Microcline and anorthoclase show albite twinning combined with a second repeated twinning almost at right angles known as pericline twinning (see Fig. 7.13). Pericline twinning is also present in plagioclases but is rarely so well developed as the albite twinning.

7.11 Calcic plagioclase (labradorite) showing characteristic broad lamellae of albite twins. × 30. Crossed nicols.

Albite twinning in plagioclase is often visible with the hand lens and is the best criterion for the distinction of plagioclase from orthoclase in hand-specimens.

Colour—usually white or pink in the hand-specimen. *Streak*—white. *Lustre*—vitreous. *H.*—6. *S.G.*—2·6.

Thin Section Characters

Colour. All the feldspars are colourless in thin section.

Form. Most often seen as sub-hedral rectangular grains or anhedral grains. Euhedral crystals are common as phenocrysts in volcanic rocks.

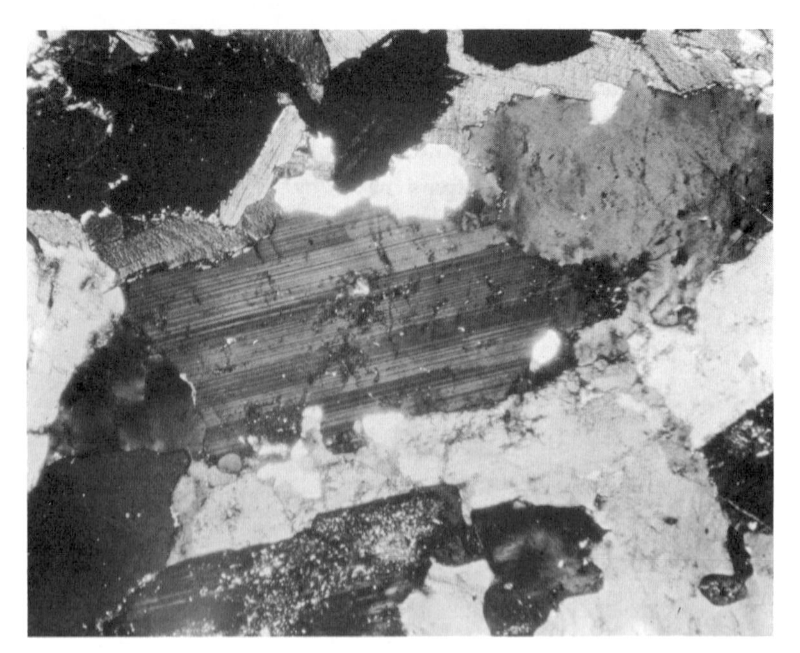

7.12 Plagioclase (oligoclase) showing the narrow albite twin lamellae characteristic of the more sodic plagioclases. ×30. Crossed nicols.

7.13 Characteristic cross-hatched twinning of microcline. ×30. Crossed nicols.

Cleavage. Usually visible in thin sections as two directions at or near 90 degrees in sections cut normal to the *a*-axis. Other sections will show no cleavage or one direction only.

Birefringence. All the feldspars have low birefringence (about 0·007–0·011) and show grey or white first-order polarization colours.

Relief. All the feldspars show low relief.

Distinguishing features. The plagioclases are distinguished from the other feldspars by the almost universal presence of albite twinning.

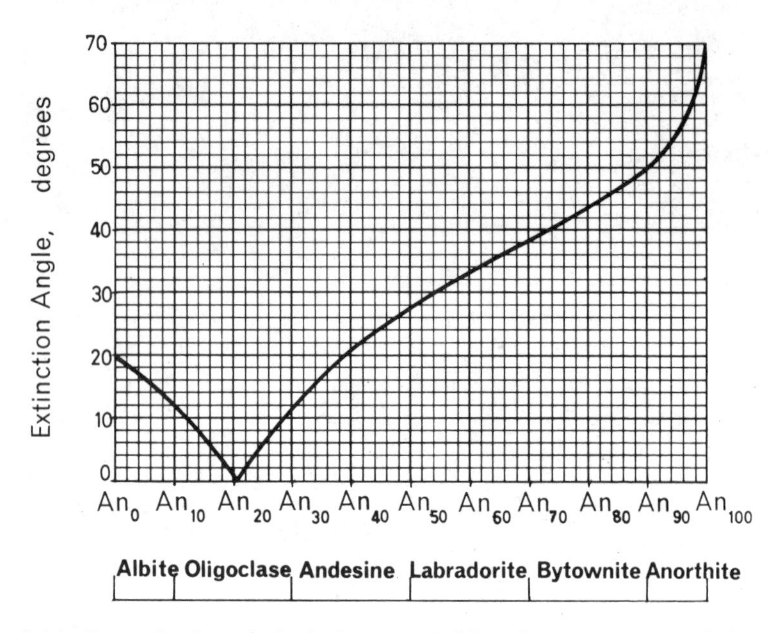

7.14 Determination of plagioclase compositions by measurement of the maximum symmetrical extinction angle in the zone normal to (010). See p. 93 for full description of the method.

Compositions within the plagioclase series are determined by the extinction angle method given on p. 93, coupled with a determination of the refractive index relative to balsam. Albite has refractive indices below balsam while the other plagioclases have refractive indices above balsam.

Orthoclase and sanidine resemble each other and have refractive indices below balsam. Orthoclase is much the more common and is found in many metamorphic and coarse grained igneous rocks. Sanidine occurs in the limited environment of volcanic igneous rocks and is usually distinguished from orthoclase by its clarity and lack of alteration.*

Microcline and anorthoclase like the other alkali feldspars have refractive indices below balsam. Both are characterized by cross-hatched twinning (see Fig. 7.13) which, however, is better developed in microcline than in anorthoclase. Microcline occurs in many metamorphic rocks, in pegmatites and in granites, whereas anorthoclase, like sanidine, is restricted to volcanic rocks.

The distinguishing features of the feldspars are summarized in Table 7.8.

Table 7.8

DISTINGUISHING CHARACTERS OF THE FELDSPARS IN THIN SECTION

	Refractive Indices below balsam	*Refractive Indices above balsam*
No lamellar twinning	ORTHOCLASE SANIDINE (volcanics)	
Lamellar twinning in one direction only	PLAGIOCLASE (Albite)	PLAGIOCLASE (Oligoclase to Anorthite)
Lamellar twinning in two directions giving a cross-hatched appearance	MICROCLINE ANORTHOCLASE (volcanics)	

Perthites. As was shown in Fig. 7.9, the solid solution between sodic and potassic feldspars is only complete in feldspars formed at high temperatures. Thus, many feldspars formed at a high temperature in both igneous and metamorphic rocks exsolve into two separate feldspars on cooling. Such feldspars have a characteristic streaked appearance due to the presence of blebs and patches of exsolved plagioclase within a host-crystal of orthoclase (see Fig. 7.15). Feldspars of this type are known as *perthites* and are extremely abundant. Un-exsolved sodi-potassic feldspars such as anorthoclase and sanidine are by contrast characteristic of rapidly cooled volcanic rocks in which there has been insufficient time for exsolution to take place.

* The certain identification of feldspars requires advanced techniques including interference figures and the determination of refractive indices.

Distinction of feldspars from quartz. The only feldspar likely to be confused with quartz is sanidine, since both minerals are characteristically clear and unaltered. Quartz, however, has refractive indices above balsam, has no cleavage, and is not twinned. In a volcanic rock containing both sanidine and quartz, however, it may not be possible to determine the identity of all the grains present because of the frequent difficulty of obtaining Becke lines on grains not on the edge of the slide.

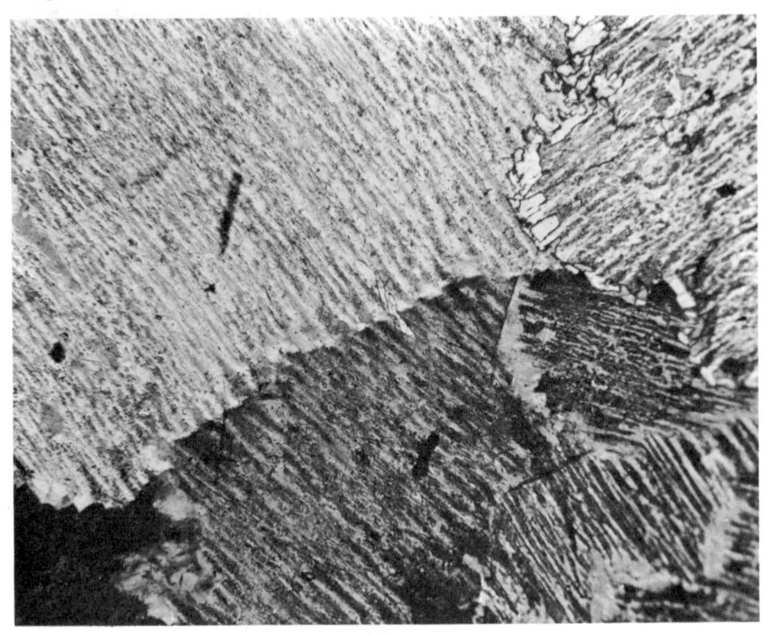

7.15 Perthitic-structure in alkali feldspar. ×50. Crossed nicols.

Occurrence. The feldspars are highly important constituents of most igneous and metamorphic rocks and many sediments. The acid and alkaline igneous rocks are characterized by alkali feldspars and calcium-poor plagioclases such as oligoclase, while more basic igneous rocks contain calcic plagioclases such as labradorite. Most types of feldspar are found in metamorphic rocks, and as detrital grains in impure sediments such as arkoses and greywackes.

THE FELDSPATHOID GROUP

The feldspathoids are an important group of rock-forming minerals which play a role similar to that of the feldspars in the undersaturated igneous rocks (see p. 152). The most important members of the group are:

Nepheline	$NaAlSiO_4$	Hexagonal
Leucite	$KAlSi_2O_6$	Pseudo-cubic
The Sodalite group	NaAl silicates with some Cl, S, and Ca	Cubic

Comparison of the formulae of nepheline and leucite with the feldspars, albite, and orthoclase, illustrates the lower silica content of the feldspathoids, since the formulae can be represented:

Nepheline	$Na_2O.Al_2O_3.2SiO_2$
Albite	$Na_2O.Al_2O_3.6SiO_2$
Leucite	$K_2O.Al_2O_3.4SiO_2$
Orthoclase	$K_2O.Al_2O_3.6SiO_2$

NEPHELINE $NaAlSiO_4$ Hexagonal

Crystallography. Crystals are stumpy hexagonal prisms terminated by the basal pinacoid.

Cleavage. Moderate, prismatic.

Colour—white or grey to pink. *Streak*—colourless. *Lustre*—greasy. Transparent to translucent. *H.*—5·5–6. *S.G.*—2·6.

Distinguishing features in hand-specimens. When well crystallized, e.g., as phenocrysts in volcanic rocks, the shape of the crystals is distinctive. Generally the greasy lustre may be used to distinguish nepheline from quartz and feldspars, though the distinction is often difficult.

Optical properties. Colour—colourless in thin sections. *Form*—euhedral crystals often found in volcanic rocks, show rectangular or hexagonal sections. In plutonic rocks nepheline is usually sub- to anhedral. *Cleavage*—the cleavage is usually visible in thin sections but tends to be somewhat irregular. It is seen in grains cut parallel to the *c*-axis. *Relief*—very low. Refractive indices are almost the same as Canada Balsam. *Birefringence*—low (0·004) giving first-order greys. *Extinction*—the rectangular sections show straight extinction; basal (hexagonal) sections are dark between crossed nicols.

Distinguishing features. Nepheline can be very difficult to identify without the use of interference figures. Confusion may arise with quartz and with alkali feldspars such as orthoclase and sanidine.

With a little experience, however, the presence of a cleavage, the straight extinction on the cleavage, and the darker grey polarization colours will all help to exclude quartz. The straight extinction on the cleavage trace also excludes the feldspars, and a Becke line test is also often useful. The refractive indices of most alkali feldspars are sufficiently far below balsam to give a clear indication of negative relief. Similarly, quartz gives a clear indication of positive relief. Nepheline may give either result but with a much less intense Becke line.

Occurrence. In undersaturated igneous rocks such as nepheline syenite, phonolite, ijolite.

LEUCITE $KAlSi_2O_6$ Pseudo-cubic

Crystallography. Commonly found as trapezohedral crystals, the form {211} (see Fig. 7.16).

Colour—white to grey. *Streak*—colourless. *Lustre*—dull to vitreous, translucent to opaque. *H.*—5·5–6. *S.G.*—2·5.

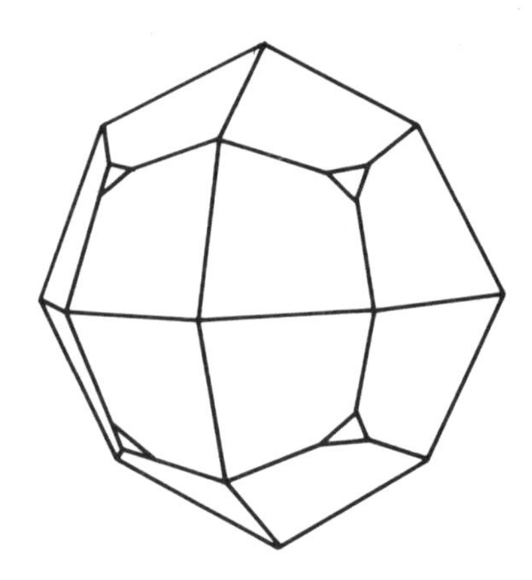

7.16 Leucite. Trapezohedral crystal modified by small octahedral faces.

Optical properties. Colour—colourless in thin sections. *Form*—the crystals characteristically show almost circular outlines (see Fig. 7.17). *Cleavage*—not usually visible in thin sections. *Relief*—moderate, negative. *Birefringence*—very low (about 0·001). Leucite crystallizes in the cubic system but subsequently inverts to a feebly birefringent tetragonal form. Polarization colours are very dark first-order greys. *Twinning*—leucite shows a characteristic polysynthetic twinning in which different sectors of the crystal are differently oriented.

Distinguishing features. The form of the crystals in hand-specimens is distinctive but they are similar to those of analcite. Leucite is usually an easy mineral to identify in thin sections because of its form and twinning. Analcite is in many respects similar but does not show twinning.

7.17 Phenocrysts of leucite in lava. Note that for clarity the lamellar twinning is somewhat over-emphasized by the contrast of this photograph. Under normal conditions of microscopy the leucite crystals would look rather darker. ×30. Crossed nicols.

Occurrence. Found both as phenocrysts and a groundmass mineral in potassium-rich volcanics poor in silica, e.g., leucite tephrite, leucitite, etc. The lavas of Vesuvius produce fine specimens.

THE SODALITE GROUP Cubic

The minerals of this group are completely isotropic and are identified by their moderately high negative relief.

THE ZEOLITE GROUP

The zeolites are a group of hydrated sodium-calcium-alumino-silicates found mainly in amygdales in lavas such as basalt. Their chief interest lies in their property of forming media for ion exchange, which finds a common application in water-softening devices.

As a group the zeolites may be identified by their occurrence and by their tendency to form good transparent or white crystals, often fibrous. In thin sections they are characterized in addition by very low refractive indices (moderate negative relief) and low to nil birefringence. Only one zeolite, analcite, is generally considered as a rock-forming mineral.

ANALCITE $NaAlSi_2O_6 . H_2O$ Cubic

Crystallography. When well crystallized, occurs in equant trapezohedral crystals similar to those of leucite. Normally found as interstitial anhedral grains in basaltic rocks.

Cleavage. Poor cubic.

Colour—usually white or greyish. *Streak*—colourless. *Lustre*—transparent to translucent. *H.*—5·5. *S.G.*—2·25.

Optical properties. *Colour*—colourless in thin sections. *Form*—usually anhedral interstitial grains. *Cleavage*—seen indistinctly as two directions at right angles in some sections. *Relief*—moderate negative. *Birefringence*—very low to nil. Highest polarization colours are very dark grey.

Distinguishing features. Difficult to distinguish from leucite in hand-specimens. In thin sections distinguished by its lack of the characteristic leucite twinning and by its interstitial occurrence.

Occurrence. As a late-stage mineral in certain undersaturated basaltic rocks, e.g., teshenite. Also abundant as an amygdale mineral in basic lavas.

THE PYROXENES

An important group of rock-forming minerals, mainly found in the igneous rocks. In composition the pyroxenes are widely variable but the following varieties are most common:

Augite (monoclinic)	$Ca(Mg,Fe)Si_2O_6$
Hypersthene (orthorhombic)	$(Mg,Fe)SiO_3$

Like the olivines the pyroxenes grade from magnesium-rich to iron-rich varieties, but may be calcium-bearing (the clinopyroxenes, monoclinic) or calcium-free (the orthopyroxenes, orthorhombic). In addition, alkaline rocks such as trachytes and syenites may contain a sodium-bearing clinopyroxene, *aegirine*.

Crystallography. All the pyroxenes are characterized by a prismatic development in which the forms {100}, {010}, and {110} are usually conspicuous. Figure 7.18 shows a typical augite crystal with hemi-pyramidal terminations.

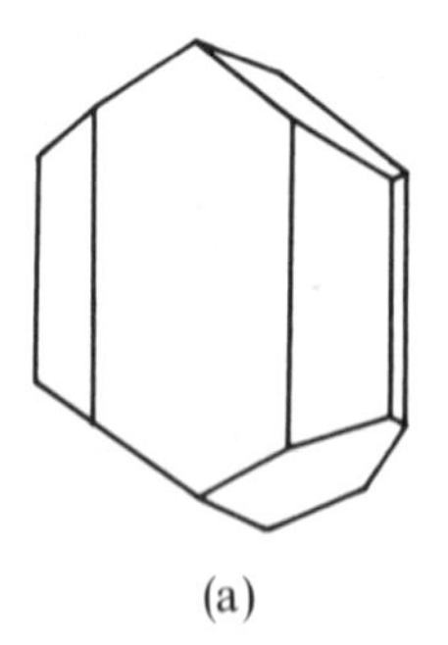

(a)

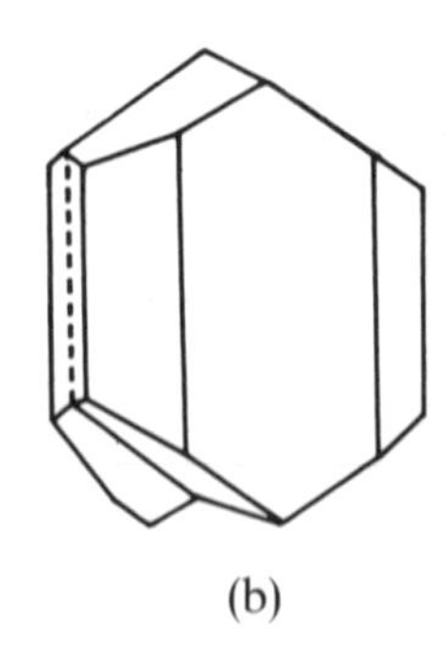

(b)

7.18 (a) Untwinned augite crystal showing prismatic forms with hemi-pyramidal ($\bar{1}11$) termination. (b) A similar crystal twinned on (100).

Cleavage. All pyroxenes have a perfect cleavage parallel to {110}, which gives two cleavage directions meeting at almost 90 degrees (see Figs. 6.3 and 7.19).

Colour—usually almost black in hand-specimens. *Streak*—white. *Lustre*—usually vitreous, may approach metallic in some specimens. Transparent to opaque. *H.*—about 6. *S.G.*—3·2–3·5.

Optical properties: general features. All pyroxenes show fairly high positive relief in thin sections and have a characteristic pattern of cleavages. Most sections show one cleavage trace but sections normal to the *c*-axis show the characteristic appearance of two cleavage traces at 90 degrees. The eight-sided nature of these sections in euhedral crystals, is also very typical.

Distinction of different pyroxenes. The *clinopyroxenes* are distinguished by their moderate birefringence (about 0·025, giving second-order colours) and their oblique extinction in sections cut approximately parallel to the *c*-axis (sections showing one cleavage trace only). The *orthopyroxenes*, such as hypersthene, have low birefringence (about 0·010, giving first-order colours) and straight extinction in most sections. Whereas most pyroxenes are almost colourless to neutral in thin sections, *aegirine* is dark yellowish green and somewhat pleochroic. Some specimens of hypersthene have a faint pink to green pleochroism.

Distinction from other minerals. In hand-specimens pyroxenes are most conveniently distinguished from amphiboles, which they otherwise closely resemble, by the possession of two cleavages at 90 degrees, as opposed to 124 degrees in the amphiboles. In thin sections the cleavage again distinguishes the two minerals providing that the

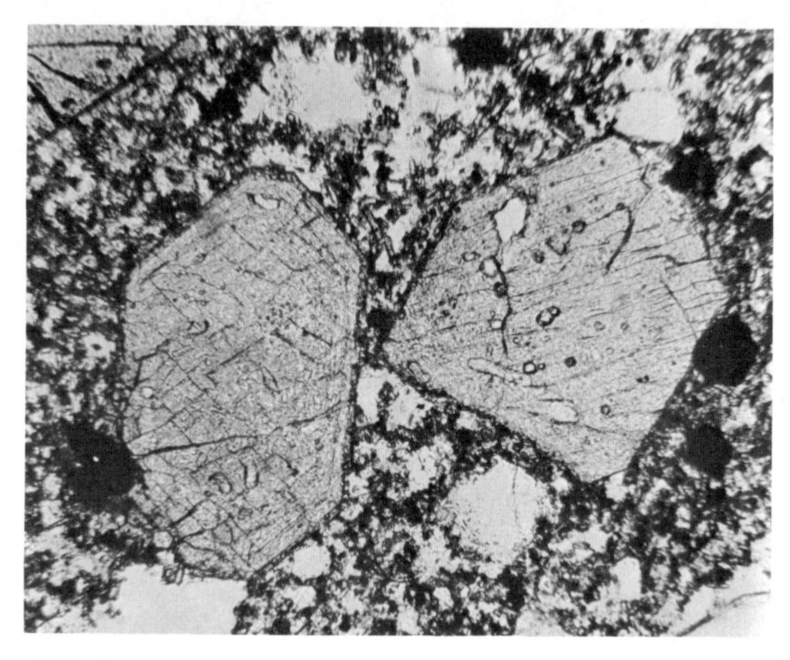

7.19 Euhedral phenocrysts of augite in a lava. ×80. Plane polarized light.

basal sections are examined. Amphiboles in addition are most commonly coloured and pleochroic and on this ground can only be confused with sodic pyroxenes such as aegirine. The extinction angle onto the trace of the cleavage in sections approximately parallel to the *c*-axis is also distinctive, being usually about 45 degrees for the clinopyroxenes and 20 degrees or less for the amphiboles. Again there may be some difficulty distinguishing aegirine from the amphiboles because it may have a low extinction angle. Examination of the cleavage should settle any doubts.

The orthopyroxenes, having low polarization colours and straight extinction, are not usually confused with the amphiboles, but may, in metamorphic rocks be difficult to distinguish from andalusite.

The well-developed pyroxene cleavage is usually sufficient to prevent confusion of the group with olivines.

Occurrence. The pyroxenes are a widely distributed mineral group but are most common in basic and ultrabasic igneous rocks where they are frequently associated with olivine. They also occur in high-grade metamorphic rocks and in marbles, and as detrital grains in immature sediments such as greywackes.

THE AMPHIBOLES

The amphiboles are an important group of rock-forming minerals in both igneous and metamorphic rocks, and are complex silicates, principally of magnesium, iron, calcium, and aluminium. Chemically they are broadly akin to the pyroxene group but differ in containing (OH). Like the pyroxenes, some varieties of amphibole found in alkaline rocks contain notable amounts of sodium.

Crystallography. Nearly all amphiboles are monoclinic and have a characteristic prismatic form (see Fig. 7.20).

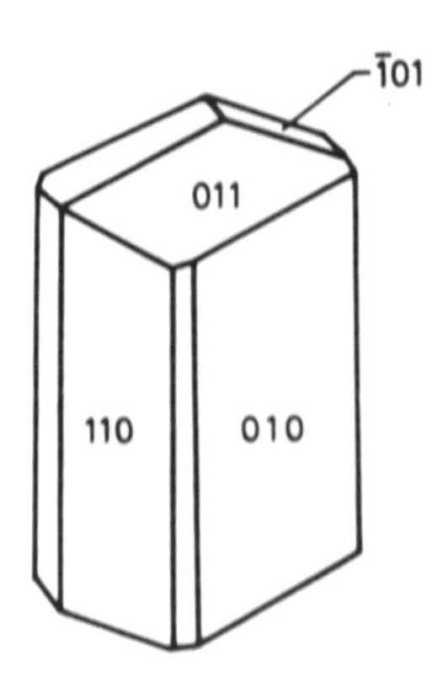

7.20 A characteristic hornblende crystal showing prismatic development and terminated by three faces, two belonging to the clinodome form (011) and the third to the hemiorthodome ($\bar{1}$01).

Cleavage. All amphiboles have good cleavages parallel to {110}, giving two directions at 124 degrees (cf. pyroxene cleavages at 90 degrees, see Fig. 6.3).

Colour—variable but usually black to dark green. *Streak*—colourless. *Lustre*—glassy. *H.*—5–6. *S.G.*—about 3.

Optical properties. Colour—colour broadly distinguishes varieties in thin sections. Common *hornblende* is strongly pleochroic in shades of brown, green, yellow, and bluish green. *Tremolite* is colourless. *Riebeckite*, a soda-amphibole, is pleochroic from almost opaque to dark inky blue and pale yellow. Another soda-amphibole, *glaucophane*, is pleochroic in shades of pale blue and violet. *Form*—elongated euhedral to sub-hedral crystals. Riebeckite often forms interstitial patches enclosing other minerals. *Cleavage*—very prominent in thin sections. Sections cut parallel to the *c*-axis show one

7.21 Hornblende in amphibolite. Note the angle between the cleavages in the basal section (centre). × 30. Plane polarized light.

cleavage trace. Basal sections (those cut normal to *c*) show two cleavages at 124 degrees (see Fig. 7.21). *Relief*—high positive. *Birefringence*—moderate (0·02–0·03). Highest polarization colours belong to the second order, but may be masked in deeply coloured specimens. *Extinction*—sections showing one cleavage trace have extinction angles of up to 25 degrees. Basal sections have symmetrical extinction.

Distinguishing features. In hand-specimens the cleavage angle of 124 degrees distinguishes amphiboles from the otherwise closely similar pyroxenes. Tourmaline can resemble amphibole but does not have a prominent cleavage.

In thin sections the cleavage is again distinctive of the group. The small extinction angle onto the cleavage helps to distinguish the amphiboles from other pleochroic minerals such as biotite and tourmaline, both of which have straight extinction. Tremolite, a colourless amphibole, is similarly distinguished from the colourless mica, muscovite. In case there should be any confusion with pyroxenes, the amphibole extinction angle, commonly of 10–20 degrees also distinguishes the group from the orthopyroxenes (straight extinction) and the clinopyroxenes (45-degree extinction).

Occurrence. Hornblende is an important constituent of many intermediate and acid igneous rocks while the soda-amphibole riebeckite is found in alkaline rocks. Tremolite occurs widely as an alteration product of other ferromagnesian minerals in igneous rocks.

Hornblende is also abundant in medium- and high-grade metamorphic rocks, especially in the amphibolites. Tremolite is found in certain marbles. Glaucophane occurs commonly in certain rocks formed in high-pressure regional metamorphism.

Amphiboles are only found rarely as detrital grains in sediments.

OLIVINE Solid solution between Mg_2SiO_4(Forsterite) and Fe_2SiO_4(Fayalite) Orthorhombic

Crystallography. Short prismatic crystals with domes and pyramids (see Fig. 7.22).

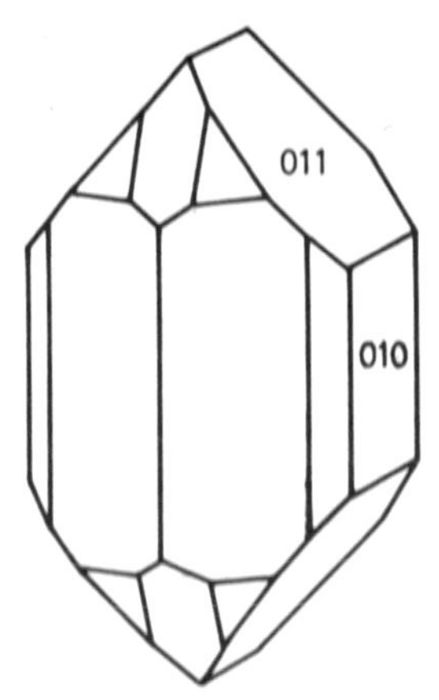

7.22 An olivine crystal showing prism, dome, and pyramid forms.

Cleavage. Not usually seen in hand specimens.

Colour—often greenish, also in shades of red, brown, and yellow. A pale green olivine of gem quality is called *peridote*. *Streak*—colourless. *Lustre*—vitreous, transparent to translucent. *Fracture*—conchoidal. *H.*—6–7. *S.G.*—3·2–4·3, depending on the composition.

Optical properties. Colour—magnesian olivines are colourless in thin section. Iron-rich varieties are yellowish brown. *Form*—often very distinctive in thin section (see Fig. 7.23). Sections of euhedral crystals are often rectangular with pointed ends. *Cleavage*—poor

7.23 Euhedral olivine crystal (centre) set in black glass in basalt. This section is cut parallel to (100) and shows the poor (010) cleavage. The shape should be compared with the olivine crystal shown in Fig. 7.22. × 30. Plane polarized light.

parallel to {010}. Not usually conspicuous in thin sections. *Fracture*—irregular curving fractures are a characteristic feature (see Fig. 7.24). *Relief*—fairly to very high, positive. *Birefringence*—high (0·03 to 0·05). Most olivines show second-order polarization colours, iron-rich olivines show third-order colours. *Extinction*—parallel to the cleavage if seen. Parallel or symmetrical with respect to crystal outline in most sections. *Alteration*—characteristically partially altered to magnetite granules distributed along irregular cracks, to serpentine minerals, and to a brown mineraloid, iddingsite.

Distinguishing features. In hand-specimens distinguished from pyroxenes and amphiboles by the lack of cleavage, conchoidal fracture, and (usually) colour. In thin sections the form, type of alteration, and straight extinction prevent confusion with any other common mineral.

Occurrence. Magnesian olivine is an important constituent of basic and ultrabasic igneous rocks. Iron-rich olivines are a relatively rare

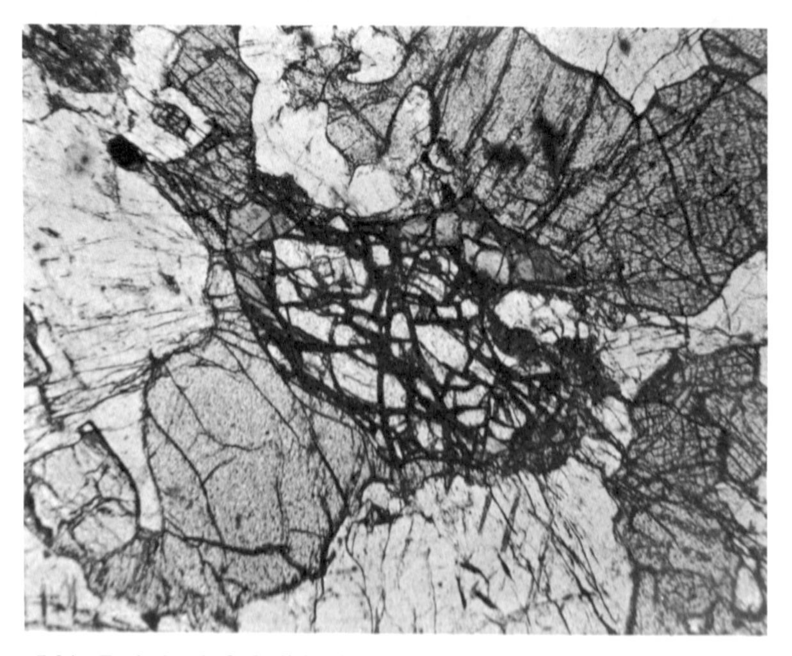

7.24 Typical anhedral olivine in gabbro, showing curving cracks filled by alteration products. Compare the appearance with that of the augite (high relief) on either side. ×30. Plane polarized light.

constituent of granites, syenites, etc. Pure magnesian olivine is found in some marbles formed by the metamorphism of impure dolomitic limestones.

GARNET Cubic

The garnets are important minerals in metamorphic rocks and form solid solutions between a number of end members, the most important of which are:

Pyrope	$Mg_3Al_2(SiO_4)_3$
Almandine	$Fe_3Al_2(SiO_4)_3$
Grossular	$Ca_3Al_2(SiO_4)_3$

Crystallography. Garnets are usually very well crystallized, the commonest forms being the rhombdodecahedron and trapezohedron (see Fig. 7.25).

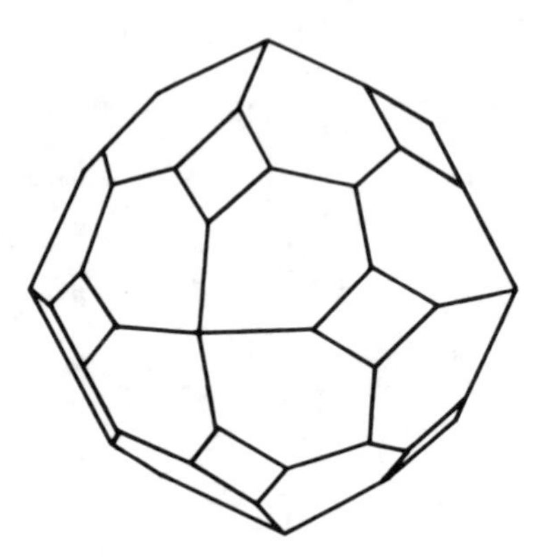

7.25 Garnet crystal showing a predominantly trapezohedral form modified by smaller rhombdodecahedral faces.

Cleavage. None.

Colour—commonly red or red brown. *Streak*—colourless. *Lustre*—glassy to resinous, transparent to opaque. *H.*—6–7·5. *S.G.*—about 4.

Optical properties. *Colour*—colourless to pale pink in thin sections. *Form*—almost circular polygonal crystal outlines (see Fig. 7.26). Often densely packed with inclusions of other minerals. *Cleavage*—none. *Relief*—high positive. *Birefringence*—nil. Remains dark between crossed nicols.

Distinguishing features. The form of the crystals is very distinctive in hand-specimens. In thin section garnet is only likely to be confused with the very much less common spinel. The occurrence in schists is typical.

Occurrence. A common mineral in medium- to high-grade metamorphic rocks. Rarely as a detrital mineral in sandstones.

7.26 Typical appearance of garnet in schist. ×50. Plane polarized light.

BERYL $Be_3Al_2Si_6O_{18}$ Hexagonal

Crystallography. Crystals common. The common form is the hexagonal prism with terminations by the basal pinacoid and pyramid (see Figs. 4.12 and 4.13).

Cleavage. Poor parallel to the basal pinacoid.

Colour—emerald green (the variety known as emerald), pale blue (aquamarine), and blue green. *Streak*—colourless. *Lustre*—vitreous, transparent to translucent. *H.*—8. *S.G.*—2·7.

Distinguishing features. Readily identified by colour, form of crystals and hardness.

Occurrence. Occasionally in granites as an accessory mineral. Usually in pegmatites and veins. Gem quality beryl, particularly emerald, is a highly-prized precious stone.

IDOCRASE $Ca_2Al_2(OH,F)Si_2O_7$ (Vesuvianite) Tetragonal

Crystallography. Crystals not uncommon, showing good prismatic form combining {100} and {110}. Terminations are usually a combination of basal pinacoid and pyramids (see Fig. 4.8).

Cleavage. Poor parallel to {100}.

Colour—dark green, brown, yellowish. *Streak*—colourless. *Lustre*—vitreous, sometimes a little resinous, transparent to translucent. *H.*—6·5. *S.G.*—3·4.

Distinguishing features. The tetragonal crystals are very distinctive.

Occurrence. Found in some marbles, notably in the ejected blocks of metamorphosed limestone thrown up by the volcano Vesuvius.

TOPAZ $Al_2F_2SiO_4$ Orthorhombic

Crystallography. Crystals common, made up of a combination of prismatic, pinacoidal, domal, and pyramidal faces. A typical topaz crystal is illustrated in Figs. 1.11 and 4.11.

Cleavage. Perfect parallel to (001).

Colour—commonly pale orange-yellow but also colourless. *Streak*—white. *Lustre*—vitreous. *H.*—8. *S.G.*—3·5.

Distinguishing features. The crystals are easily recognizable when well formed and there is little chance of confusion with other exceptionally hard minerals such as beryl and corundum.

Occurrence. Found in certain granitic igneous rocks and in pegmatites and veins often associated with cassiterite and tourmaline. Topaz has some value as a gem stone.

ANDALUSITE Al_2SiO_5 Orthorhombic

Crystallography. Crystals are common, the usual habit being prismatic with an almost square cross-section (form {110}). Crystals are terminated by a basal pinacoid only.

Cleavage. Poor parallel to {110}.

Colour—grey, to purplish and pinkish red. Surfaces are often altered to white mica. *Streak*—white. *Lustre*—vitreous, transparent to opaque. *H.*—7·5, though altered surfaces may be softer. *S.G.*—3·2.

Optical properties. Colour—colourless or occasionally very pale pink (pleochroic) in thin section. *Form*—elongated crystals or columnar aggregates with rectangular or square cross-sections. Included opaque material in the variety known as *chiastolite* may form a sort of maltese cross in cross-sections (see Fig. 7.27). *Cleavage*—moderately prominent in thin sections. One cleavage trace is visible in sections cut near-parallel to the *c*-axis and two intersecting at nearly 90 degrees are seen in cross-sections. *Relief*—moderate, positive. *Birefringence*—low (about 0·009) giving first-order colours a little higher than those of quartz. *Extinction*—parallel to crystal outlines and cleavage in most sections. Symmetrical in cross-sections.

Distinguishing features. Readily identified in hand-specimens because of the form of the crystals. In thin sections can be confused with hypersthene though the shapes of cross-sections are different and andalusite has a lower relief.

Occurrence. Andalusite is found in metamorphosed rocks of argillaceous composition.

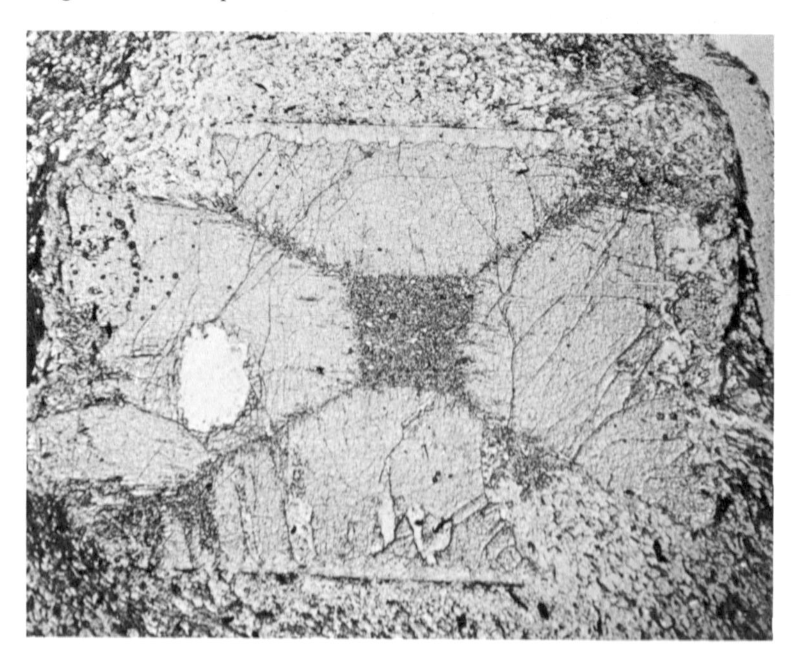

7.27 Chiastolite showing cruciform arrangement of inclusions. × 30. Plane polarized light.

KYANITE Al_2SiO_5 Triclinic

Crystallography. Usually in elongated bladed crystals embedded in schists.

Cleavage. Perfect parallel to (100) and less perfect parallel to (010) and (001).

Colour—usually bluish, frequently mottled. A very distinctive appearance. *Streak*—white. *Lustre*—vitreous, transparent to translucent. *H.*—varies considerably depending on the direction of scratching, e.g., 5 along the direction of elongation of the crystals and 7 across this. *S.G.*—3·7.

Optical properties. Colour—colourless to pale blue. *Form*—broad elongated sections parallel to (100) and narrower sections parallel to (010) and (001). *Cleavage*—very conspicuous in thin sections. Many sections show two cleavage traces approximately at right angles (see Fig. 7.28). *Relief*—high positive. *Birefringence*—moderate (0·016). Interference colours up to first-order red. Most sections show greys, yellow or orange. *Extinction*—depending on the section varies from almost straight to an extinction angle of 30 degrees onto the cleavage.

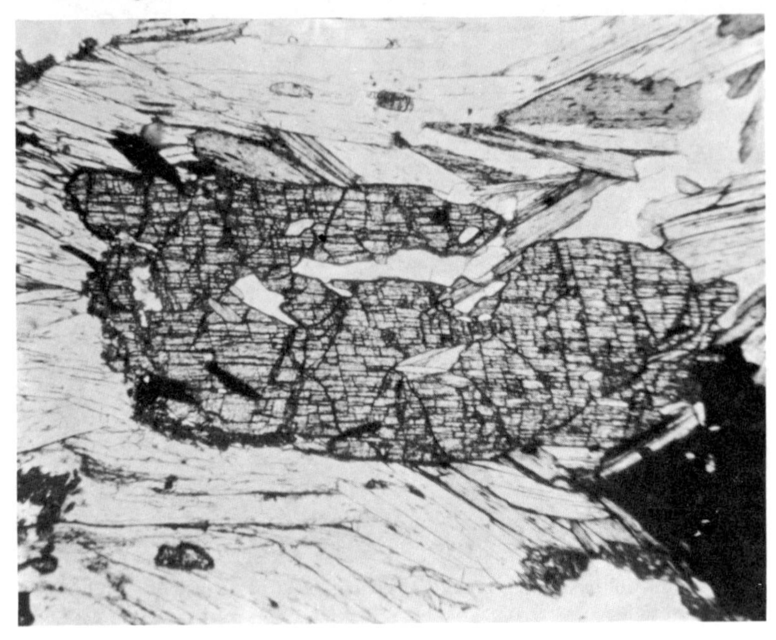

7.28 Kyanite in schist, showing high relief and cleavage in two directions. × 30. Plane polarized light.

Distinguishing features. Easily identified in hand-specimens by the bladed form, characteristic mottled blue colour, and hardness. In thin sections it might be confused with orthopyroxene or andalusite but both of these are orthorhombic, having straight extinction in prismatic sections and symmetrical extinction in basal sections. Kyanite also has a conspicuously higher relief than andalusite and hypersthene.

Occurrence. In schists and gneisses derived by metamorphism of argillaceous rocks. Commonly associated with garnet and staurolite.

SILLIMANITE Al_2SiO_5 Orthorhombic

Crystallography. Usually long acicular crystals or fibrous aggregates.

Cleavage. Perfect parallel to (010).

Colour—brown, greenish, pale grey. *Streak*—colourless. *Lustre*—vitreous, transparent to translucent. *H.*—6–7. *S.G.*—3·3.

Optical properties. *Colour*—colourless in thin sections. *Form*—small to very small acicular crystals and fibrous masses. Crystals are often bent. *Cleavage*—not normally conspicuous in thin sections. *Relief*—high positive relief. *Birefringence*—moderate (0·02), comparable to clinopyroxenes, giving second-order colours. *Extinction*—the elongated crystals have straight extinction.

Distinguishing features. The fibrous and acicular form of the crystals, embedded in rock, is distinctive.

Occurrence. A constituent of high-grade schists and gneisses.

TOURMALINE Complex silicate of aluminium, boron, iron, magnesium, and sodium Trigonal

Crystallography. Good crystals commonly found. The form is prismatic to acicular, the crystals being three-sided in cross-section. The termination is usually a flat rhombohedron (see Fig. 4.19).

Cleavage. Poor rhombohedral.

Colour—usually black, occasionally green, blue, or red. There is frequently some zonation of the colours about the long axis of the crystal. *Streak*—colourless. *Lustre*—vitreous, transparent to opaque. *H.*—7. *S.G.*—3·0.

Distinguishing features. The black acicular crystals may be mistaken for hornblende but the latter has a good prismatic cleavage. In good crystals the trigonal form of tourmaline is distinctive.

Occurrence. Mainly a pegmatite mineral and an accessory mineral in granite.

EPIDOTE $Ca_2(Fe,Al)_3(OH)(SiO_4)_3$ Monoclinic

Crystallography. Crystals common though usually small. The common habit is acicular, the elongation being along the *b*-crystallographic axis.

Cleavage. Perfect parallel to (001), not usually observed in hand-specimens because of the small size of the crystals.

Colour—a distinctive green colour known as pistacchio. *Streak*—white. *Lustre*—vitreous, transparent to translucent. *H.*—6·5. *S.G.*—3·4.

Optical properties. Colour—colourless to very pale yellow in thin sections, often variable within a single crystal. *Form*—often in small elongated crystals. *Cleavage*—parallel to the length of the crystals since the elongation is in *b*. *Relief.* very high, positive. *Birefringence*—moderate (0·014) to strong (0·045) depending on the iron content. Colours range up to third order in the iron-rich varieties. Sections showing mid first-order colours have a slight, anomalous, bluish or greenish yellow tinge to the greyish polarization colour. *Extinction*—straight extinction in most elongate sections.

Distinguishing features. In hand-specimen the form of the crystals and the colour are distinctive. In thin sections the pale yellow colour in plane polarized light, the bright interference colours and the straight extinction are sufficient to distinguish epidote from most minerals. The anomalous first-order polarization colours are also very distinctive with experience.

Occurrence. A very widely distributed, though rarely abundant mineral. Found as a detrital mineral in sediments and as a common constituent of many metamorphic rocks. A widespread secondary product in igneous rocks.

STAUROLITE $2Al_2SiO_5 . Fe(OH)_2$ Orthorhombic

Crystallography. Commonly well crystallized in stumpy prismatic crystals showing the forms {110}, {010}, and {001} (see Fig. 7.29). Interpenetration twins are common (see Fig. 4.24).

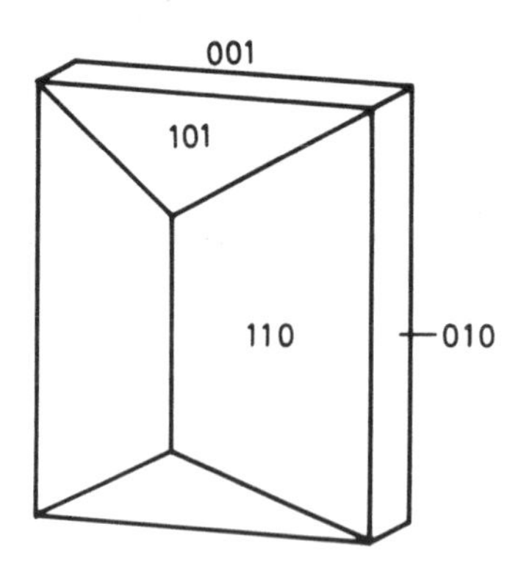

7.29 Untwinned staurolite crystal.

Cleavage. Poor parallel to {010}.

Colour—dark brownish. *Streak*—colourless. *Lustre*—glassy to dull, usually opaque in hand-specimens. *H.*—7–7·5. *S.G.*—3·7.

Optical properties. *Colour*—pale yellow in thin sections with distinct pleochroism. *Form*—usually in well-formed crystals (see above), often large. *Cleavage*—inconspicuous parallel to {010}. *Relief*—high positive. *Birefringence*—moderate (0·012). Polarization colours mainly greys and yellows of the first order. *Extinction*—parallel in most sections, symmetrical in cross-sections.

7.30 Staurolite in schist. The high relief, inclusions, parallel-sided shape, and lack of cleavage, are all characteristic. × 30. Plane polarized light.

Distinguishing features. In hand-specimens the form of the crystals and their twinning is distinctive. In thin sections the colour and pleochroism make staurolite an easy mineral to recognize.

Occurrence. In schists and gneisses derived from the metamorphism of iron-rich argillaceous sediments. Frequently associated with garnet and kyanite.

THE MICA GROUP

The micas are an important group of rock-forming minerals characterized by their perfect basal cleavage and the ease with which they can be cleaved into very thin elastic plates. The two most important members of the group are *muscovite*, a hydrated silicate of potassium and aluminium, and *biotite*, a hydrated silicate of potassium, magnesium, and iron.

MUSCOVITE Monoclinic

Crystallography. Platy six-sided crystals of pseudo-hexagonal appearance, made up of the forms {001}, {010}, and {110} (cf. Fig. 7.31). The angle β is near 90 degrees. Also disseminated and in scaly aggregates.

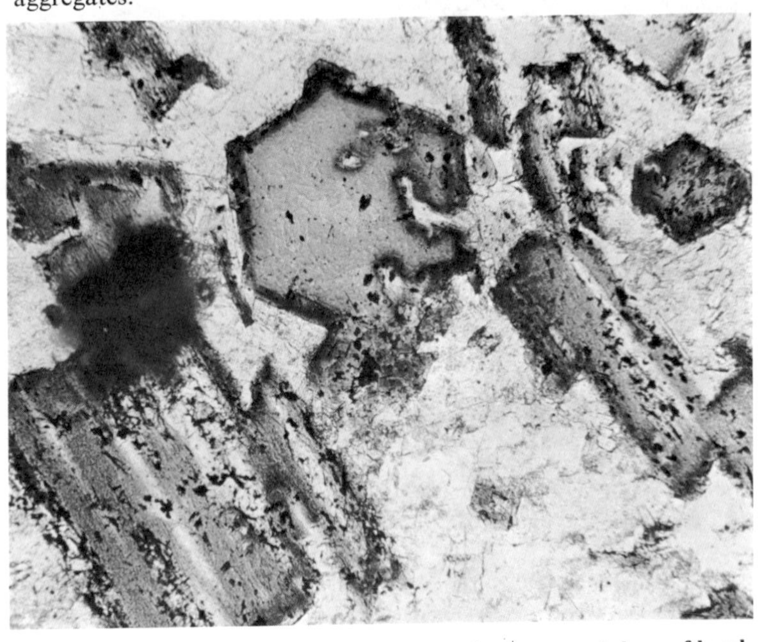

7.31 Euhedral crystals of biotite in lava, showing hexagonal shape of basal section, rectangular shape of other sections. $\times$30. Plane polarized light.

Cleavage. Perfect parallel to the basal pinacoid. The cleavage flakes are elastic when bent.

Colour—white, colourless, often gives a silvery impression. *Streak*—colourless. *Lustre*—pearly. Often shows iridescence due to incipient parting of cleavage planes internally. *H*.—2–2·5. *S.G.*—2·8.

Optical properties. Colour—colourless in thin sections. *Form*—usually in tabular crystals which appear mainly as elongated sections in the slide. Basal sections are equidimensional. *Cleavage*—very prominent in one direction. Not seen in basal sections. *Relief*—varies from low positive to moderate positive as the crystals are rotated in plane polarized light. *Birefringence*—high (about 0·04). Polarization colours range up to the top of the second order. Basal sections show first-order colours. *Extinction*—parallel to the cleavage trace in elongated sections.

Distinguishing features. In hand-specimens the micaceous form together with the lack of colour, or silvery appearance, are characteristic. In thin sections the bright polarization colours immediately bring the mineral to the attention. The good cleavage and the straight extinction are then sufficient for identification. Basal sections are, however, difficult to spot as they have low polarization colours and show no cleavage.

Occurrence. As large crystals in pegmatites and as a common constituent of many schists. Also found in granites and as a widespread alteration product of feldspars in other igneous rocks. In the latter case it is usually extremely finely divided and is known as *sericite*.

BIOTITE

Biotite is very similar to muscovite but is dark brown to black in hand-specimens. It shows a very brilliant lustre on cleavage surfaces.

In thin sections biotite exhibits intense pleochroism, having its maximum absorption when the trace of the cleavage is parallel to the vibration direction of the lower nicol. The pleochroism is commonly from brown to pale yellowish brown, though red-brown or green biotites are also found. Like muscovite, biotite has high birefringence and straight extinction. The polarization colours may be masked by the absorption of the crystals. Basal sections of biotite show no pleochroism, a strong absorption, no cleavage trace, and no birefringence.

Biotite is easily distinguished from most minerals by its micaceous form (see Fig. 7.31) and intense pleochroism. It may resemble hornblende in thin sections but is distinguished by its straight extinction (cf. most hornblendes show extinction angles of about 20 degrees).

Occurrence. Biotite is an extremely common mineral, being an important constituent of many schists and gneisses. In igneous rocks it is rarely very abundant but is very common, particularly in acid and intermediate rocks such as granites, syenites, diorites, etc.

CHLORITE GROUP

The term chlorite covers a group of hydrated magnesium iron aluminium silicates of generally micaceous appearance. A closely related group of similar properties is termed *serpentine*.

Crystallography. The chlorites are monoclinic and some are pseudo-hexagonal like the micas. They are, however, rarely well crystallized enough to enable any determination of the morphology to be made.

Cleavage. Perfect basal cleavage similar to that of the micas.

Colour—shades of green. *Streak*—colourless. *Lustre*—vitreous to pearly. Transparent to opaque. *H.*—about 2. Can be scratched with the finger nail. *S.G.*—2·6–3.

Optical properties. Colour—deep green to pale green or colourless in thin section. *Form*—tabular crystals showing elongated sections. Also in vein-like and radiating aggregates. Often replacing other minerals, particularly ferromagnesian minerals. *Cleavage*—elongate sections show a prominent cleavage parallel to the length. *Relief*—rather low positive. An iron-rich chlorite, chamosite, found in an oolitic form in sedimentary ironstones has a somewhat higher moderate positive relief. *Birefringence*—generally weak to nil. Polarization colours are often anomalous, e.g., inky blue, dark brown, purplish brown. *Extinction*—parallel to the cleavage trace in most chlorites.

Distinguishing features. Chlorite-rich rocks such as chlorite schist may sometimes be identified by their green colour and slightly greasy feel. In thin sections chlorites are distinguished from green biotite or hornblende by their lower relief and weak birefringence. Specimens showing anomalous polarization colours are usually readily identifiable.

Occurrence. Chlorite is an important constituent of low-grade metamorphic rocks. Chamosite is an important economic mineral as a constituent of sedimentary iron ores. In igneous rocks chlorites are very common alteration products of other minerals and also occur as an amygdale-filling material.

ZIRCON $ZrSiO_4$ Tetragonal

Crystallography. Prismatic tetragonal crystals with pyramidal terminations are common but are usually very small (see Fig. 4.7).

Cleavage. Poor parallel to the faces of the prism {110}.

Colour—colourless, yellowish, greenish, grey, brown. *Streak*—colourless. *Lustre*—adamantine, transparent to opaque. *H.*—7·5. *S.G.*—4·7.

Optical properties. Colour—colourless in thin sections. *Form*—very small prismatic or equidimensional grains in igneous and metamorphic rocks. May be found as more rounded grains in sediments. *Cleavage*—not conspicuous. *Relief*—extremely high positive relief. *Birefringence*—very high (0·06). Third- and fourth-order colours are typical. *Extinction*—straight extinction in elongated sections.

Distinguishing features. The occurrence as small grains with extremely high relief is characteristic. Sphene has a different crystal form (see Fig. 6.5) and shows high-order white polarization colours but may sometimes be difficult to distinguish from zircon in very small grains. Apatite is distinguished from zircon by lower relief and low birefringence.

Occurrence. Zircon is an extremely common though rarely abundant accessory mineral in acid and intermediate igneous rocks, where it not infrequently is found as inclusions in biotite. The zircons in biotites are often surrounded by dark zones known as pleochroic haloes. Large zircon crystals may be found in pegmatites.

Zircon is a resistant mineral and may appear as detrital grains in sediments.

SPHENE $CaTiSiO_5$ Monoclinic

Crystallography. Wedge-shaped or flat tablet-shaped crystals are characteristic. Twinning is common and gives sharp re-entrant angles.

Cleavage. Moderate, prismatic.

Colour—commonly brownish or yellowish. Also green, grey. *Streak*—colourless. *Lustre*—slightly resinous, transparent to opaque. *H.*—5. *S.G.*—3·5.

Optical properties. Colour—colourless to neutral or slightly brownish in thin sections. *Form*—usually rather small crystals. An elongated diamond-shaped section is characteristic (see Fig. 6.5). Also in small rounded droplets. *Cleavage*—not normally very noticeable in thin sections. *Relief*—extremely high positive relief. *Birefringence*—extremely high (0·12). High-order white is the characteristic polarization colour. Sphene is the only common mineral which has a birefringence comparable to that of the carbonates, calcite, and dolomite. *Extinction*—the diamond-shaped sections have symmetrical extinction. Sphene does not always extinguish completely on rotation of the stage.

Distinguishing features. The high polarization colours distinguish sphene from other minerals except the carbonates. The high relief and the crystal form, however, should prevent any confusion. Small granules of sphene may be difficult to identify but in some cases their association with ilmenite in the form of reaction rims is indicative.

Occurrence. Sphene is a very widespread accessory mineral in igneous rocks, particularly granites and syenites. It is also common in metamorphic rocks of many types.

8. Igneous rocks

Principles of Classification

Rocks are classified and given names so that information about their constitution can be conveyed in a condensed fashion. The criteria of classification can be genetic, mineralogical or chemical, and systems based on each of these features, or combinations of them, have been widely used for igneous rocks. All have their particular value but none can be completely satisfactory because rocks, unlike, for example, plants and animals, do not usually fall into natural and distinct species but are continuously variable. Any classificatory scheme must therefore consist of divisions which are to some extent arbitrarily selected, and it unfortunately follows that rock names can be used differently by different authorities.

A simple classification based on mineralogy is probably the most useful for everyday purposes and a substantial amount of agreement is found between the many schemes proposed, providing too much elaboration and sub-division is avoided. The particular advantages of such a system are firstly that the mineralogical composition of a rock is often relatively easy to determine using a microscope, and secondly that the mineralogy to a large extent reflects the bulk chemical composition of the rock. Difficulties are, however, encountered when the rocks become too fine-grained to allow accurate mineral determination, and are extreme when the rocks contain notable amounts of glass.

For the sake of completeness, and in order to provide a general picture of the whole spectrum of igneous rocks it has been thought necessary in this chapter to include a number of families of rock-types and certain individual rock-types which are either rather rare or difficult to identify without the use of advanced techniques. Principal amongst these are the rocks belonging to the Alkaline Ultrabasic family, and the nepheline-bearing members of the

Feldspathoidal syenite family. Many elementary teaching collections will not therefore contain samples of these rocks, but the authors feel that it will do no harm for the elementary student to be aware of their existence and to know where they fit in the general pattern.

Chemical Compositions of Igneous Rocks

Table 8.1 shows the approximate average chemical compositions of some important igneous rock families and Fig. 8.1 shows how they may be laid out in diagrammatic form when the total of the alkali oxides ($Na_2O + K_2O$) is plotted against silica (SiO_2). The name of

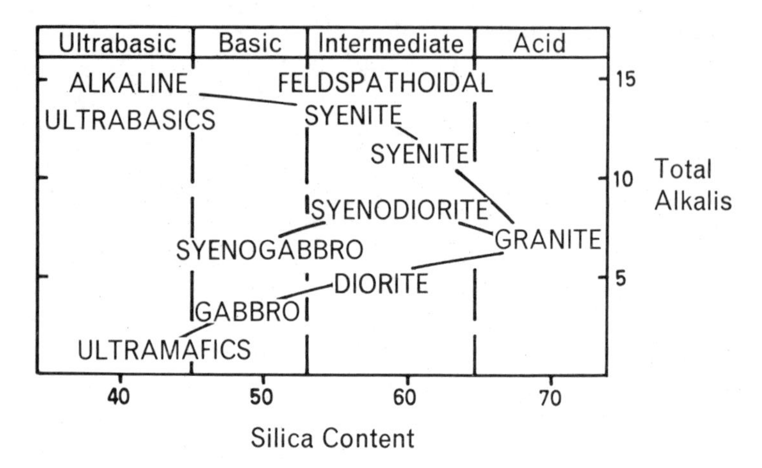

8.1 Alkali-silica diagram showing the positions of the main families of igneous rocks. The heavy lines indicate the series illustrated in Fig. 8.2. Note that no genetic relationships are implied by the three series chosen.

the family in each case refers to the *coarse grained* or *plutonic* (see p. 157) representative of each group. Inspection of Table 8.1 will show that a rise in silica and alkalis is generally accompanied by a fall in MgO, CaO, and iron oxides, while Al_2O_3, the remaining important constituent, is rather less variable with the exception that it falls away sharply in the ultramafic rocks which are composed largely of alumina-free or alumina-poor minerals such as olivine, pyroxene, and hornblende.

Apart from the division into families, the igneous rocks are commonly referred to as *acid, intermediate, basic*, or *ultrabasic* depending on their silica content (see Fig. 8.1), and are also divided into the *calc-alkaline* and *alkaline* suites. The term calc-alkaline refers to the

Table 8.1

APPROXIMATE AVERAGE COMPOSITIONS OF SOME COMMON IGNEOUS ROCKS
(after Daly)

Family:	*Granite*		*Diorite*	*Gabbro*	*Ultramafic*		*Syeno-diorite*	*Syenite*	*Feldspathoidal Syenite*	*Alkaline Ultrabasic*	*Syeno-gabbro*
Rock type:	*Granite*	*Grano-diorite*	*Diorite*	*Gabbro*	*Picrite*	*Dunite*	*Monzonite*	*Syenite*	*Nepheline Syenite*	*Ijolite*	*Essexite*
SiO_2	70·8	65·7	57·6	49·0	43·1	41·7	56·7	62·5	55·4	43·2	49·3
TiO_2	0·4	0·6	0·9	1·0	0·9	—	1·1	0·6	0·9	1·6	1·9
Al_2O_3	14·6	16·1	16·9	18·2	9·9	0·9	17·1	17·6	20·2	19·1	18·2
Fe_2O_3	1·6	1·8	3·2	3·2	5·5	2·9	3·0	2·1	3·4	3·9	4·4
FeO	1·8	2·7	4·5	6·0	9·3	5·7	4·1	2·7	2·2	4·9	5·7
MgO	0·9	1·9	4·2	7·6	20·8	47·7	3·3	0·9	0·9	3·2	4·1
CaO	2·0	4·5	6·8	11·2	8·4	0·7	6·6	2·3	2·5	10·6	9·0
Na_2O	3·5	3·7	3·4	2·6	1·3	0·1	3·7	5·9	8·4	9·7	4·4
K_2O	4·2	2·8	2·2	0·9	0·4	—	3·8	5·2	5·5	2·3	2·3

gabbro-diorite-granodiorite series in which plagioclases (calcium-sodium feldspars) play an important role. The term alkaline is usually reserved for the syenite, nepheline syenite, and alkaline ultrabasic groups where alkali feldspars and feldspathoids are important, while the syenogabbros and syenodiorites may be referred to as mildly alkaline. Granites may be referred to as alkaline or calc-alkaline, depending on their individual characters.

Mineralogical Compositions

The bulk variations in chemistry can be related to the mineralogy of the rocks when the chemical compositions of the more important rock-forming minerals are considered (see Table 8.2).

Thus the low content of both SiO_2 and alkalis in the ultramafic rocks implies the presence in quantity of the ferromagnesian (also referred to as 'mafic') minerals, which are low in silica and alkalis, e.g., olivine, pyroxene, and hornblende. Increase of silica and alkalis leads to the gabbros, with the incoming of calcic plagioclase, a mineral with rather higher SiO_2 and alkalis. Further increase leads via the diorites to the granites where the high silica minerals, quartz, and the alkali feldspars, are abundant, and calcium-bearing plagioclase and ferromagnesian minerals are subordinate. The high content of alkali feldspars, of course, reflects the fairly high alkali contents of the rocks.

The granites grade into the syenites by the increase of alkalis (increase of alkali feldspar) and the decrease of SiO_2 (disappearance of quartz). If this trend* is continued, the feldspathoids (minerals with higher alkali contents than the alkali feldspars but lower SiO_2 contents) make their appearance in the feldspathoidal syenite family. Extreme reduction in the amount of SiO_2 while maintaining high alkalis leads eventually to alkaline ultrabasic rocks such as ijolite, a rare pyroxene-nepheline rock quite devoid of feldspars.

The ferromagnesian minerals in general become less abundant with the combined rise of alkalis and SiO_2. The percentage of ferromagnesian minerals in any rock is referred to as the 'colour index' and this is low for the granites, syenites, and feldspathoidal syenites (e.g., 25 per cent or less), moderate for the diorites, gabbros, syenogabbros, and many of the alkaline ultrabasics (e.g., 40–50 per cent) and is high for the ultramafic rocks (up to 100 per cent).

Figure 8.2 summarizes the essential mineralogical variation in the

* The term 'trend' is not used here in any genetic sense, i.e., to imply that the rocks referred to are actually produced in this way.

Table 8.2

APPROXIMATE CHEMICAL COMPOSITIONS OF COMMON MINERALS OF IGNEOUS ROCKS
(in weight %)

		SiO_2	Al_2O_3	MgO + FeO	CaO	$Na_2O + K_2O$
Ferromagnesian minerals or 'Coloured silicates'	olivine[a]	40	—	60	—	—
	augite[b]	50	3	23	20	—
	hornblende[b]	40	10	30	12	trace
	biotite[b]	36	15	30	1	10
Plagioclase Series	calcic plagioclase[c]	54	29	—	12	5
	sodic plagioclase (albite)	68	20	—	—	12
Alkali feldspars	potash feldspar (orthoclase, microcline etc.)	65	18	—	—	17
Feldspathoids	nepheline	42	36	—	—	22
	leucite	55	23	—	—	22
Silica minerals	quartz etc.	100	—	—	—	—

[a] Composition of the common olivine of basic igneous rocks, containing about 25% of Fe_2SiO_4 and 75% of Mg_2SiO_4.
[b] In addition to the constituents shown, augite, hornblende, and biotite contain significant amounts of TiO_2 and Fe_2O_3. Hornblende and biotite also usually contain about 2% of water.
[c] Composition of the common plagioclase of basic igneous rocks, containing about 60% of $CaAl_2Si_2O_8$ and 40% of $NaAlSi_3O_8$.

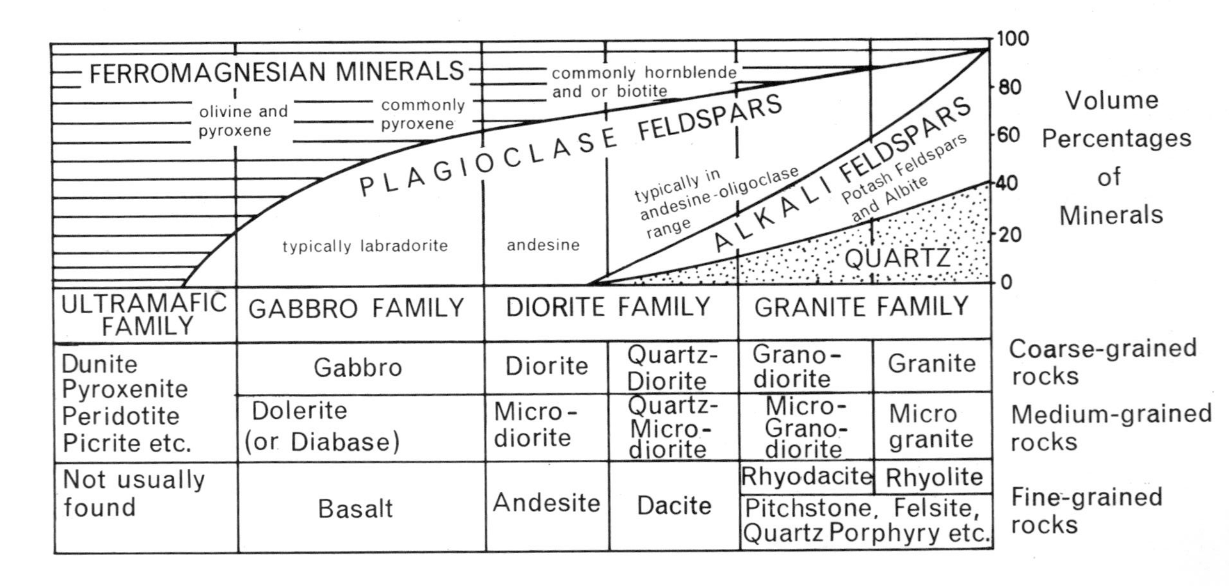

8.2(a) Mineral compositions and classification of rocks in the granite-diorite-gabbro-ultramafic range.

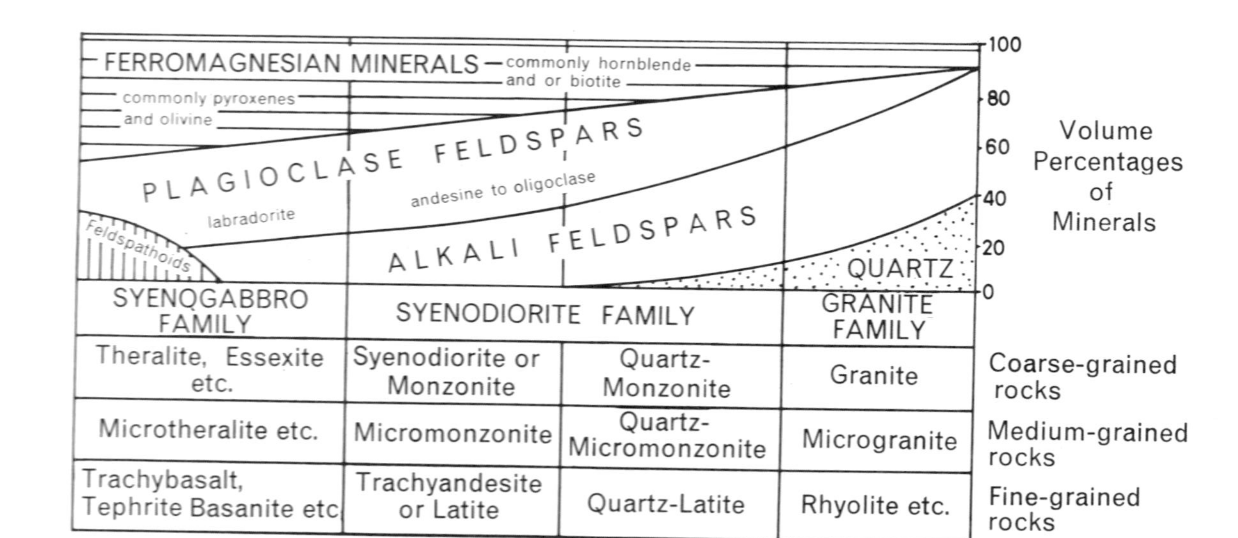

8.2(b) Mineral compositions and classifications of the rocks in the granite-syenodiorite-syenogabbro range.

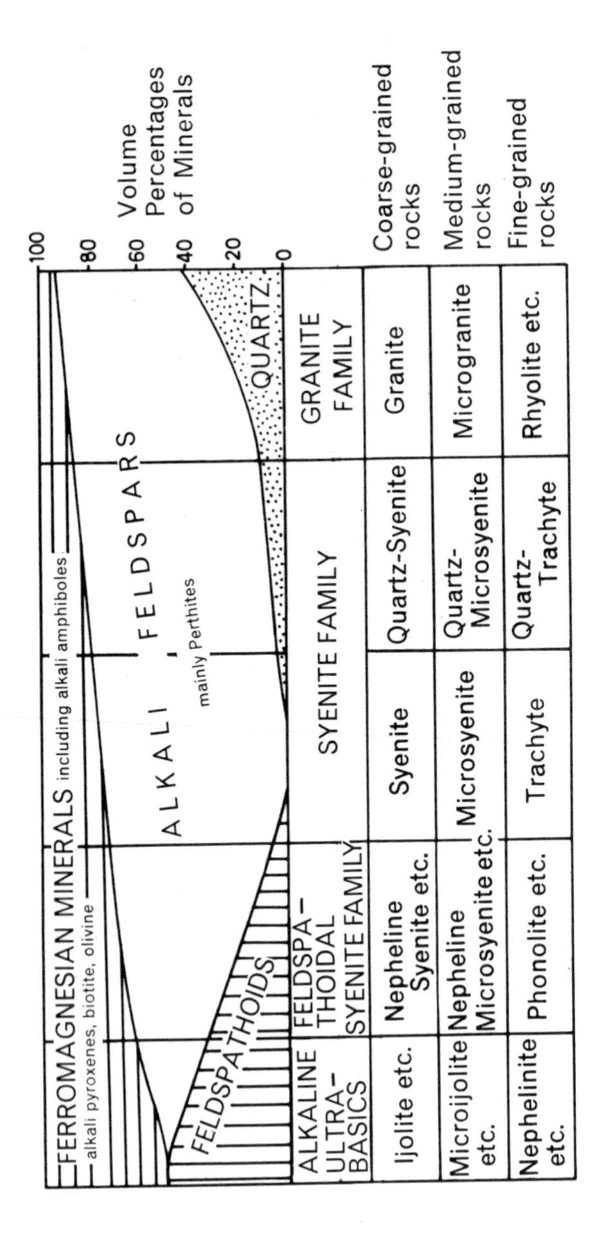

8.2(c) Mineral compositions and classifications of the rocks in the granite-syenite-feldspathoidal syenite-alkaline ultrabasic range.

igneous rocks as a whole. It is useful to visualize these diagrams in their relative positions in Fig. 8.1 and to imagine that the three diagrams (all of which take granite as a common point) can be readily linked by additional diagrams illustrating the diorite-syenodiorite-syenite series, the gabbro-syenogabbro-feldspathoidal syenite series, etc.

It is important to note that the term alkali feldspar includes all the potash feldspars together with albite, the sodic end-member of the plagioclase series. The volume percentage of plagioclase feldspars in Fig. 8.2 thus refers only to members of the plagioclase series containing more than 10 per cent (molecular) of the calcium end-member, anorthite. If this distinction were not made, rocks which were essentially similar, but differed in being highly sodic on one hand (albite-rich) and highly potassic on the other (orthoclase-rich), would fall into quite different groups. It is clearly desirable to group them together as alkali-rich rocks rather than to separate them because they contain different alkali metals.

Grain Size and Crystallization of Igneous Rocks

Igneous rocks may be further classified on the basis of their grain size into *coarse-grained* (e.g., average grain size > 5 mm), *medium-grained* (average grain size 5 mm–1 mm) and *fine-grained* (average grain size < 1 mm) although there is little unanimity on the precise dimensional limits. This sub-division is roughly equivalent to the more genetic classification into deep-seated or *plutonic* rocks, the rocks of minor intrusions (called *hypabyssal* rocks), and the extrusive *volcanic* rocks. It is perhaps preferable to use the former system in deciding on nomenclature because the necessary information is derived entirely from the specimen of rock itself. It is occasionally possible, for example, to find two identical rocks, one in a dyke (a hypabyssal rock), the other in a lava flow (a volcanic rock). It would seem more sensible to use the grain size classification and give them both the same name than to use the genetic system and call them by different names. The latter system is, however, deeply ingrained in geological usage and has the advantage of additionally conveying a limited amount of information about the field-occurrence of the rock concerned.

The coarse-grained rocks have, in general, crystallized relatively slowly at some depth within the crust, while the volcanic rocks have to a large extent crystallized more rapidly at or near the surface. This sometimes results in mineralogical as well as grain-size differences between coarse-grained and fine-grained rocks of the same bulk chemical compositions. The general principles governing the crystallization of silicate melts were expressed by Bowen in what he termed the *reaction principle.* From the study of rocks themselves and also from experimental work, it is evident that many minerals which begin to precipitate from a melt during cooling subsequently react with the remaining liquid to form a new mineral phase or an altered version of the existing phase. Thus, olivine, commonly the first phase to appear on the cooling of basic magma, may react with the liquid at a lower temperature to produce pyroxene, while itself being partly or totally resorbed. Similarly, at a still lower temperature the pyroxene may react with the liquid to produce hornblende and at a lower temperature again, the hornblende may react to give biotite. These are examples of what Bowen termed *discontinuous* reactions.

In contrast, a *continuous* reaction takes place between the liquid and minerals which form solid solutions. The first plagioclase crystals to appear, for example, in basic magmas are rich in the anorthite (calcium) end-member. As the temperature falls a progressively more and more albite (sodium)-rich plagioclase will be precipitated and, if equilibrium between crystals and liquid is maintained, the existing crystals will be made over to the more sodic composition appropriate to the lower temperature.

The reactions referred to depend, however, on the maintenance of equilibrium between the crystals and the liquid. Continuous reactions, for example, depend on a rate of cooling which is slow enough to allow the crystals to adjust their compositions to the demands of falling temperature. A rate of cooling faster than this results in a crystal being *normally zoned*, that is to say the crystal changes gradually in composition from centre to margin. This feature is very commonly seen in plagioclases, which tend to be more calcic centrally and more sodic marginally. Zoning may also be of an oscillatory nature (see Fig. 8.3).

Discontinuous reactions may be suppressed almost entirely in cases of rapid cooling which will therefore give rise to rocks containing minerals which are higher members of the reaction series compared with rocks produced from the same magma by slow cooling. Thus in a slowly cooled (plutonic) rock such as granite the common ferromagnesian minerals are biotite and hornblende, while the extrusive, rapidly-cooled, equivalent (rhyolite) quite commonly contains pyroxene and, occasionally, olivine. Thus, one general effect of the reaction of minerals with liquid is to give differing mineralogies to rocks of the same composition. A more important drawback to mineralogical classification is, however, caused by the failure of reactions in rapidly cooled volcanic rocks, since the early formed minerals are most commonly more basic (in the sense that they contain more CaO, MgO, and Fe oxides) than the rock as a whole. If, as often happens in volcanic rocks, a period of slow crystallization (corresponding with the magma's slow rise to the surface) is followed by rapid chilling on extrusion, a porphyritic rock (see p. 162) is formed which has rather 'basic' phenocrysts and a less 'basic' groundmass.

8.3 Plagioclase in trachybasalt, showing oscillatory zoning. × 30. Crossed nicols.

The groundmass mineralogy of such a rock is frequently indeterminate, and the phenocrysts by themselves may give a false impression of the bulk composition of the rock, generally making it appear more basic than it actually is.

Classification Tables

Table 8.3 gives a simple classificatory scheme for igneous rocks in hand-specimen, while a fuller mineralogical classification suitable for thin-section work is incorporated in Fig. 8.2. As far as possible

Table 8.3

HAND-SPECIMEN IDENTIFICATION OF THE MORE COMMON IGNEOUS ROCKS

	Generally light-coloured rocks				Generally dark-coloured rocks
	Content of Ferromagnesian minerals typically 5–10%		typically 30–50%		typically 75% or more
	Quartz rare or absent	*Quartz conspicuous*	*Hornblende conspicuous*	*Pyroxene or Pyroxene and Olivine conspicuous*	Pyroxene only—PYROXENITE Olivine only—DUNITE
COARSE GRAINED	QUARTZ SYENITE SYENITE	GRANITE (in the broad sense) including: Granodiorite Adamellite	DIORITE	GABBRO (in the broad sense) including: Norite Troctolite Olivine gabbro, etc.	Pyroxene and Olivine— PERIDOTITE Serpentine minerals— SERPENTINITE
MEDIUM GRAINED	MICROSYENITE	MICROGRANITE	MICRODIORITE	DOLERITE or DIABASE	The rocks above are rarely found in fine-grained varieties.
FINE GRAINED	TRACHYTE RHYOLITE Generally pale-coloured rocks. Rhyolites may contain visible quartz as phenocrysts. Non-porphyritic rhyolites and trachytes are difficult to distinguish. Note that glassy rhyolites (obsidian pitch-stone) are dark in colour.		ANDESITE BASALT Generally medium- to dark-coloured rocks. If phenocrysts of olivine are present the rock is probably basaltic. Hornblende phenocrysts suggest andesite. Andesites on the average are a little paler than basalts but there are many exceptions to this.		

the medium-grained rocks are given the same name as the coarse grained rock of the same composition but prefixed by 'micro-'.

It is common practice to sub-divide further by prefixing rock-names with the names of characteristic minerals of the rock. One may distinguish, for example, biotite-granite and hornblende-granite, nepheline-syenite and sodalite syenite, and augite-andesite, hypersthene-andesite, and hornblende-andesite. The minerals used in this way must not be essential to the preliminary classification of the rock since the latter are implied by the name given, e.g., terms such as quartz-granite, augite-gabbro would never be used. It is, however, useful to refer to porphyritic rocks using the suffix '-phyric' after the name of the phenocryst mineral, e.g., pyroxene-phyric basalt, or olivine-plagioclase-phyric basalt.

In Fig. 8.2 the problem of different usages of rock names arises. Where alternative names are given (e.g., dolerite *or* diabase) this is largely a result of different usages in Great Britain and North America. The North American two-fold division of the acid rocks into granodiorite and granite is, for reasons of simplicity, preferred to the British three-fold division, granodiorite-adamellite-alkali granite.

Textures of Igneous Rocks

A large number of terms has been coined to describe textural features of igneous rocks but relatively few are needed in normal use. The most useful are listed below:

(a) *Terms Describing the Degree of Crystallinity:*

holocrystalline—entirely composed of crystals.

glassy—strictly used to mean completely glassy (non-crystalline) rocks. Loosely used to refer to rocks predominantly composed of glass.

(b) *Terms Referring to Shapes of Mineral Grains:*

euhedral (*idiomorphic*)—showing perfect or near perfect crystal form (see Fig. 7.23).

anhedral (*allotriomorphic*)—showing no regular crystal form.

subhedral—a useful term to describe grains which show a recognizable but imperfect crystal form.

Depending on the habit of the crystals concerned, terms such as equant (equidimensional), platy, tabular, acicular, etc., may be used in the same way as they are used for minerals in hand-specimens.

(c) *Terms Referring to the Relationships between Different Grains:*

ophitic—a texture common in dolerites (diabases) illustrated in Fig. 8.4. The plagioclase crystals are euhedral and the augite fills the angular interstices, the same crystal of augite enclosing or partly enclosing several plagioclase crystals.

8.4 Ophitic texture in gabbro. A single large augite crystal (medium grey) fills the interstices between and partially encloses a number of well-formed plagioclase crystals (black, white, and grey). × 30. Crossed nicols.

interstitial—a term used for a mineral which has crystallized late, filling up the available space between earlier crystals.

intergrowths—between two minerals; occasionally found. The commonest of these is the fine-grained regular or lobate intergrowth of quartz and alkali feldspar which is called *micropegmatite* or *granophyric* intergrowth (see Fig. 8.5).

(d) *General Textural Terms:*

porphyritic (see Fig. 8.6)—the common texture of volcanic rocks. Early formed crystals, often euhedral and formed before a lava reaches the surface, are set in the much finer aggregate of minerals resulting from the rapid cooling of the lava after extrusion. The larger, early crystals are referred to as *phenocrysts* and the fine-grained material as the *groundmass*. The texture is less common in coarser grained rocks. A rock not showing the contrast in grain size between phenocrysts and groundmass is referred to as *non-porphyritic*.

8.5 Granophyric texture in granophyre. A crystal of alkali feldspar (grey) is intergrown with a quartz crystal (white). × 100. Crossed nicols.

trachytic—the term used when the small feldspar laths in volcanic groundmasses are oriented by the flow of the magma during cooling (see Fig. 8.6).

granular—used to describe rocks whose constituent crystals are anhedral or sub-hedral, and equidimensional. The common texture of granitic rocks is of this type and the term *granitic* is often used in a textural sense to describe it.

basaltic—the typical groundmass texture of basalts, it consists of small elongated plagioclase laths randomly oriented and associated with small equant grains of

8.6 A euhedral phenocryst of alkali feldspar, showing Carlsbad twinning, is set in a groundmass of smaller feldspar laths showing a trachytic structure. $\times 30$. Crossed nicols.

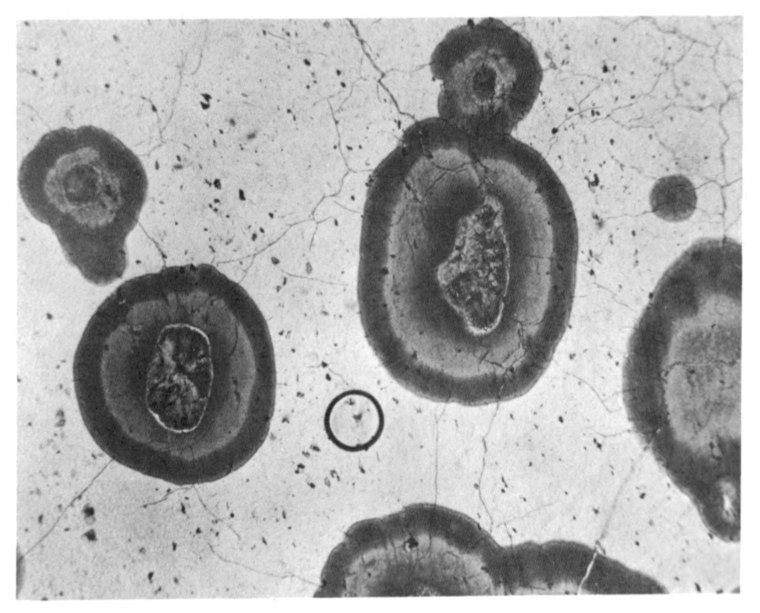

8.7 Spherulites growing in glassy pitchstone. The small dark ring is caused by a bubble in the Canada Balsam (a not uncommon feature of thin sections).

pyroxene or pyroxene and olivine often with some interstitial glass.

spherulitic—used for rocks containing radiating aggregates of crystallites (microscopic crystals) (see Figs. 8.7 and 8.8). A common texture in glassy rocks which are partially devitrified, i.e., in glasses which have begun to crystallize in the solid state after the original consolidation of the rock.

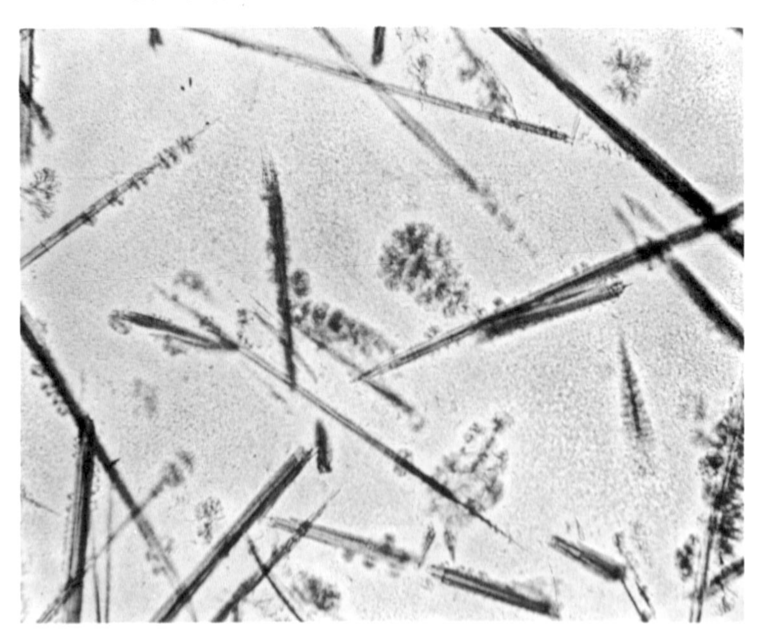

8.8 Feathery microlites of clinopyroxene in pitchstone.

Other structures include *vesicles*—empty holes representing original gas bubbles, *amygdales*—vesicles filled by later-crystallizing minerals (see Fig. 8.9), *xenoliths*—fragments of foreign rock incorporated in the magma, and *xenocrysts*—accidentally included crystals from a foreign source.

The Petrographic Description of an Igneous Rock

In describing a thin section it is desirable to follow a set pattern so that nothing important is omitted and a logical layout is achieved.

Section 1 should be a brief and comprehensive description of the mineralogy and texture of the rock, summarizing the essential features. A reader can thus see at a glance whether the rock under discussion is going to be of interest to him and can decide whether or not to read the further details.

Section 2 should be a description of the individual minerals taken one by one in the following order (a) phenocrysts, describing the ferromagnesian minerals first, then the feldspars, then the quartz or feldspathoids (b) groundmass minerals, in the order ferromagnesian minerals, feldspars, quartz or feldspathoids, and finally accessories.

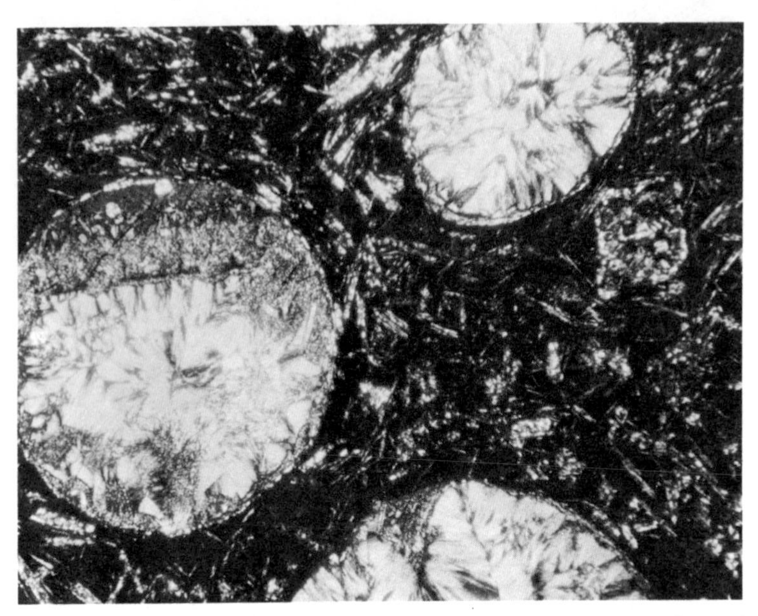

8.9 Amygdales filled by chlorite, in mainly-glassy basalt. ×30. Crossed nicols.

A mineral appearing both as phenocrysts and in the groundmass is described under both headings.

In describing a mineral, only those features which are inherently variable should be mentioned, e.g., size and shape of grains, the amount of the mineral present, the colour and colour-range of any pleochroism, the presence or absence of twinning, degree of alteration, etc. The reader does not want to know that quartz has first-order grey polarization colours or that hornblende has two cleavages at 124 degrees. These are invariable characters. A full description of all properties is only necessary when a mineral cannot be identified because this may aid its identification subsequently.

Section 3 can be reserved for any discussion of the rock which may seem relevant.

It is most important that an estimate of the volume proportions of each mineral should be made in order to classify the rock. Accurate estimates may be made using mechanical devices such as the automatic point-counter but these techniques are time-consuming and inappropriate to an elementary course. The student must normally learn to make a visual estimate of mineral proportions which may become quite accurate with practice. A convenient technique with a fine-grained rock is to consider a given field of view with a microscope and to imagine that all the visible grains of a given mineral are collected together in one clump. The size of the imagined clump may then be compared with, for example, the size of one quadrant of the field which represents 25 per cent of the *area*. It may then be reasonably inferred that a clump of this size represents 25 per cent of the *volume* of the rock. Estimates of the proportions of each mineral should be made for a number of different fields of view and the results averaged.

The system outlined above becomes difficult to use when the grain size of the rock becomes so large that only a few grains are visible in the field of view at one time. In this case better results can be obtained by examining the section without the microscope, using a hand lens. This will work well for minerals which are easy to identify, e.g., grains of biotite in a thin section of granite, or grains of plagioclase in a gabbro, but will fail in some cases where identification is uncertain, e.g., estimating the proportions of two feldspars. In the latter situation the microscopic method must be used despite its difficulties.

Having assessed the relative mineral proportions and the texture of the rock, a name can be assigned and should be added at the top of the description together with the locality. See the following brief example:

Quartz-trachyte

The rock is strongly porphyritic, consisting of large euhedral sanidine phenocrysts set in a fine-grained groundmass consisting largely of alkali feldspar laths with some interstitial quartz and hornblende.

Phenocrysts

The *sanidines* are extremely well formed, showing good crystal outlines. The crystals appear as somewhat elongated rectangles up to 5 mm in length and nearly all show simple twinning and are quite clear and unaltered. They make up about 15 per cent of the rock.

Groundmass

The only identifiable ferromagnesian mineral occurs as very small sub-hedral prisms showing strong pleochroism from pale yellow to very dark brown. It is probably hornblende and makes up about 1 per cent or less of the rock.

Alkali feldspar (? sanidine) makes up most of the groundmass and occurs as small clear laths showing simple twinning and a conspicuous flow orientation. It makes up about 75 per cent of the rock.

Quartz is present as interstitial patches and makes up perhaps 5 per cent of the rock.

Accessories include quite abundant scattered granules of an opaque ore mineral and occasional tiny crystals of probable sphene.

More or less detail may be included depending on the time available and the purpose for which the description is required.

DESCRIPTIONS OF THE FAMILIES

The Granite Family

Acid igneous rocks containing more than 10 per cent of quartz. They may be sub-divided into *granites* (in the strict sense) in which alkali feldspars (orthoclase, perthites, microcline, and albite) predominate over calcium-bearing plagioclase feldspars (usually oligoclase or andesine), and *granodiorites* where the plagioclase predominates over the alkali feldspars. The feldspars collectively make up the largest proportion of the rock and the content of ferromagnesian minerals is correspondingly low. The latter consist usually of biotite or hornblende but pyroxenes and fayalitic olivine are occasionally present. Some granites contain notable amounts of the white mica, muscovite. Accessory minerals commonly include apatite, sphene, and zircon.

The medium-grained equivalents of these rocks, usually found in small intrusions, are best referred to as *microgranite* and *microgranodiorite*. Pyroxene and olivine are somewhat more common in them than in the coarser rocks but biotite and hornblende are usual.

The fine-grained rocks of the granite family have received a variety of names based partly on texture and mode of occurrence as well as on mineralogy. The extrusive volcanic equivalents of granite and granodiorite are known as *rhyolite* and *rhyodacite*. These are usually porphyritic rocks with phenocrysts of feldspar, quartz, biotite, hornblende, pyroxene or olivine set in a very fine-grained groundmass usually somewhat banded. Varieties with prominent feldspar and quartz phenocrysts visible in the hand-specimen are known as *quartz-porphyries* and are found as extrusive and intrusive rocks. The almost completely glassy varieties are known as *pitchstone* (pitchy lustre in hand-specimen) and *obsidian* (vitreous lustre in hand-specimen). The term *felsite* is a very loosely used field term for any fine-grained crystalline acid rocks. *Granophyre* refers to medium to fine-grained rocks in which a regular intergrowth of quartz and alkali feldspar is formed (see Fig. 8.5).

Relationships. With decreasing quartz, granitic rocks grade into quartz-syenites, quartz-monzonites, and quartz-diorites (tonalites) depending whether the alkali feldspars are dominant, approximately equal in quantity to the plagioclase feldspars, or subordinate to the plagioclase feldspars.

The Diorite Family

Diorites are intermediate igneous rocks consisting essentially of intermediate plagioclase (ideally andesine) making up somewhat more than half of the rock, together with hornblende, and, commonly, biotite. Quartz and alkali-feldspar may be present in small amounts.

The coarse-grained diorites are not particularly common rocks but the fine-grained version, andesite, is a very important extrusive type. Andesites are usually porphyritic and contain phenocrysts of plagioclase, characteristically showing oscillatory zoning, together with phenocrysts of ferromagnesian minerals such as hornblende, biotite, hypersthene, and augite. The groundmasses are often too fine-grained for accurate identification but are largely composed of plagioclase microlites with some glass. Lavas somewhat similar to andesites but containing essential quartz phenocrysts are referred to as *dacites*.

Relationships. With increasing quartz, diorites grade into the granodiorites and with increasing alkali-feldspar into the monzonites (syenodiorites). As the plagioclase becomes more basic (labradorite), diorites grade into gabbros though this characteristically involves also a change of the ferromagnesian minerals from hornblende-biotite to pyroxene-olivine.

The Gabbro Family

The ideal gabbro consists of plagioclase (labradorite), making up slightly more than half the rock, together with augite and a little iron ore. The ophitic or sub-ophitic texture is common. Varieties include *olivine-gabbro* where olivine accompanies the augite, and the two augite-poor types, *troctolite* (olivine + plagioclase) and *norite* (orthopyroxene + plagioclase). Plagioclase-rich gabbros grade into *anorthosites* in which the coloured silicates are very subordinate in amount.

Microgabbros (medium-grained size) are usually referred to as *dolerite* (Gt. Britain) or *diabase* (N. America).

The fine-grained extrusive equivalents of the gabbros are termed *basalts* and are the most abundant of all lava types. Phenocrysts of plagioclase (labradorite), olivine (magnesian), and augite, or a combination of these, are common, and groundmasses consist of small granules of augite, with or without olivine, plagioclase laths, granules of ore and, commonly, some interstitial glass.

Relationships. With the incoming of alkali feldspars or feldspathoids the gabbros grade into the syenogabbros, and, with an increase of the ferromagnesian content, into the ultramafic rocks via feldspar-poor gabbroic rocks which may be termed *melagabbros*. A basaltic equivalent of melagabbro is *picrite-basalt*, a basalt unusually rich in olivine phenocrysts.

The Ultramafic Rocks

The term ultramafic is used for rocks consisting largely of coloured silicates. Many of these are in the chemical sense ultrabasic since they have low silica contents. Others such as the pyroxenites have silica contents little lower than many gabbros and are chemically speaking still basic rather than ultrabasic. It is noteworthy that the ultramafic rocks are found only with exceptional rarity in fine-grained or extrusive varieties. This is because they are mainly formed by the accumulation of crystals of ferromagnesian minerals precipitated from more normal basic magmas, not from the consolidation of magmas which are themselves exceptionally rich in iron and magnesium.

Common varieties are:

dunite—rocks mainly composed of olivine
peridotite—rocks composed of pyroxene and olivine
picrite—like peridotite but including some feldspar, or analcite. Commonly medium grained.
pyroxenite—rocks mainly composed of pyroxene
serpentinite—serpentine rocks formed by the alteration of dunite, peridotite, etc.

The Syenogabbro Family

This term covers a wide variety of basic rocks with alkaline affinities. The rocks are generally gabbroidal to basaltic in appearance but contain alkali-feldspar or feldspathoids (or the related mineral analcite) in addition to basic plagioclase. Some examples are:

theralite—a gabbroic rock with additional nepheline
essexite—a gabbroic rock with nepheline and alkali feldspar
teschenite—a gabbroic rock with analcite
basanite—olivine basalt with nepheline
tephrite—basalt with nepheline
leucite-tephrite—basalt with leucite.

The term *trachybasalt* is loosely used for basaltic rocks showing alkaline affinities in having a less basic plagioclase than normal, sometimes with interstitial analcite, alkali feldspar, etc.

The Monzonite (Syenodiorite) Family

The coarse-grained *monzonites* are intermediate in mineralogy between the syenites and diorites, typically having a fairly low colour index, the major part of the rock being made up of approximately equal amounts of plagioclase (oligoclase-andesine) and alkali feldspars. Hornblende and biotite are the common coloured silicates. The volcanic equivalents, the *trachyandesites* are commonly porphyritic rocks like andesites but having a trachyte-like (alkali feldspar-rich) groundmass. Quartz-bearing trachyandesites are termed *quartz-latites*.
Relationships. Monzonites grade into syenites with a decrease of plagioclase and into diorites with a decrease of alkali feldspar. With the incoming of quartz they grade through quartz-monzonites into granites and granodiorites. With an increased colour index and a more basic plagioclase they are transitional into the syenogabbros.

The Syenite Family

Members of this family are characterized by a low colour index and a high content of alkali feldspar. Quartz or feldspathoids may be present in accessory amounts and there may be a small amount of intermediate plagioclase. The ferromagnesian minerals are very variable and may include biotite, hornblende, alkali amphiboles such as riebeckite, the alkali pyroxene, aegirine, and the iron-rich olivine, fayalite. The alkali feldspar is frequently a strongly perthitic type.

Volcanic members of the family include the *trachytes* and quartz-trachytes, both usually porphyritic rocks with phenocrysts of alkali feldspar (commonly sanidine) and a groundmass of flow-oriented alkali feldspar laths.

Relationships. With the incoming of more than 10 per cent of quartz the syenites grade into the granites; with the appearance of more than accessory amounts of feldspathoids (usually nepheline) they grade into the feldspathoidal syenites and, with an increase of plagioclase, into the monzonites.

The Feldspathoidal Syenites

These rocks are akin to the syenites but are characterized by the presence of minerals of the feldspathoid group in addition to alkali feldspar. Nepheline is the commonest feldspathoid and gives rise to the rock-type nepheline-syenite. Other varieties such as the sodalite-syenites are rare. As in other alkaline rocks alkali-bearing ferro-magnesian minerals such as aegirine and the alkali amphiboles are common.

Volcanic members of the family are represented by the *phonolites* (nepheline-bearing trachytes) and the *leucite phonolites* where leucite accompanies or takes the place of nepheline. Other varieties may contain phenocrysts of minerals of the sodalite group.

The Alkaline Ultrabasic Family

This is a poorly defined group of rather rare, highly alkaline, low-silica, rocks of which the coarse grained *ijolite*, an aegirine-nepheline rock with colour index of about 50 per cent, forms a suitable example. The volcanic equivalent of this rock is *nephelinite*; and an example of a potassic alkaline ultrabasic lava is afforded by *leucitite*, a pyroxene-leucite rock.

Relationships. With decrease of alkalis, the alkaline ultrabasic rocks grade into the more normal ultrabasic types by the elimination of feldspathoids. With increasing alkali feldspar content they grade into feldspathoidal syenites and with the appearance of basic plagioclase into syenogabbros.

Genetic Aspects of Igneous Rocks

The study of the origins and genetic relationships of igneous rocks can very rarely be pursued without the aid of chemical analyses, and we have therefore so far concentrated on descriptive aspects of these rocks. In contrast with the sedimentary rocks which are the subject of the following chapter, individual specimens of igneous rocks rarely provide important information regarding their mode of origin. It is only by the detailed study of series of related rocks that such information may be obtained. It is important, however, that even in introductory courses the student should have some idea of the genetic processes involved so that the observations he makes may have some significance. The average thin section provides a maze of detail, amongst which some factors may be of genetic importance, and others purely irrelevant.

All igneous rocks, by definition, have been formed by consolidation of molten or partly molten material termed *magma*.

The majority of volcanic rocks, if we exclude the andesites of mountain belts and certain rhyolites, are formed by the fractional crystallization of basic to ultrabasic magmas formed within the earth's mantle, at depths of 100 km or more. During their ascent to the surface such magmas are continually subjected to the abstraction of heat, with the resultant precipitation of a variety of solid phases (minerals). Since the precipitated crystals are almost always more dense than the magma, they sink relative to the liquid and in many cases are effectively removed from it. The continuation of this process during magmatic ascent causes the composition of the residual liquid to change. This process, termed *crystal fractionation*, accounts for a substantial proportion of the variation in igneous rocks. From one initial starting composition a great variety of products can be obtained depending very largely on the rate at which the magma ascends relative to the rate at which heat is extracted from it. The earlier stages of fractionation usually involve the removal from the liquid of large quantities of ferromagnesian minerals such as olivine and pyroxene, later joined by calcium-rich plagioclase, and the liquids are thus progressively enriched in alkalis, and depleted in magnesium, calcium, and iron. Thus one of the principal factors affecting rock-variation is the extent to which the magma has followed this line of evolution at the time of eruption or intrusion and final solidification.

There is a second important factor, however, namely the branching of the evolutionary line at various stages. Depending on interacting variables such as rate of magma movement, rate of heat loss, and pressure conditions, very similar basaltic magmas can give rise to a variety of fractionation products. One basaltic magma may fractionate with silica enrichment towards a *rhyolitic* end product, while another may undergo no significant enrichment in silica and give rise to *phonolitic* end products. There are in fact a great number of possible fractionation paths passing through basalt and heading via trachybasalts and andesites towards phonolitic, trachytic, and rhyolitic residua. Intermediate types such as trachyandesite may of course be formed *en route*, and plutonic equivalents of all these types may be found as intrusions. When considering this part of magmatic evolution, however, we are largely concerned with the final stages of the history of the magma during its approach to the surface of the earth. Normally the least evolved rocks we see at the surface are the basalts we have spoken of as being parental to many other volcanics, but it should be remembered that the basalts themselves have already undergone a considerable degree of evolution before they approach the surface.

It is possible to tell what mineral phases have been involved in fractionation by examining the phenocryst assemblages in porphyritic rocks. Most phenocrysts in lavas, for example, form before the lava is erupted. At the time of eruption, intense cooling (quenching) causes the liquid part of the magma to solidify as a fine grained groundmass, while the solid phases which were already present are preserved as phenocrysts. These phenocrysts, obviously, are not the actual crystals which have been removed from the magma and caused fractionation, but they are representatives of the mineral species responsible for fractionation immediately before the eruption. Thus by collecting a series of rocks at different stages of evolution it is possible to reconstruct the fractionation process over a wide range of magma composition. The practical point to be stressed here is that this type of study can only be carried out by a careful assessment of the phenocrysts of the rocks concerned.

The crystals which sink in the magma are known by the name of *cumulus* materials. Under certain special circumstances the materials may be observed as rocks, which are genetically termed *cumulates*. Cumulates are most commonly seen in intrusive bodies, particularly intrusions of gabbro, which represent former subterranean magma chambers now exposed by erosion. In large sills, for example, the crystal fractionation process may often be studied in detail. The presence of a floor in such a magma chamber allows the cumulus material to collect as a 'sediment' towards the base of the intrusion, while rocks such as syenite or microgranite are found patchily towards the top and represent the solidification of the final residual magma. The cumulates often have a distinctive texture in which the precipitated mineral phases (e.g., olivine and plagioclase) are euhedral, and frequently aligned in a laminar structure akin to the bedding in sedimentary rocks.

Cumulus materials are also brought up in the magma from depth as nodules composed of peridotite, pyroxenite, gabbro, and similar rock-types.

We have given some indication of the flexibility of the crystallization process in producing variation in rock-types via the branching evolutionary tree of residual liquids on one hand, and the corresponding, though more rarely seen, variety of cumulates on the other. We excluded andesites and certain rhyolites earlier, not because they are not subjected to the same processes of fractional crystallization, but because they may not originate from the primitive basic magmas parental to the other rocks discussed.

Many rhyolitic and granitic magmas are thought to be formed within the crust of the earth rather than within the mantle, by the partial melting of crustal rocks. Often the agent of melting may be a copious flow of basaltic magma from beneath, so that provinces of associated acid and basic rocks are formed, devoid of intermediate varieties. It is also suggested that when mixing of the acid and basic magmas does occur, in zones of mountain building for example, the resultant magma is andesitic. There is, however, little consensus of opinion on the question of andesites and it is alternatively suggested that the presence of water within the upper mantle or crust in mountain-building zones is sufficient to produce primitive magmas which give rise to andesite rather than the usual basaltic products.

Finally we may give further consideration to that most voluminous of all plutonic rocks, granite. Granites and related rocks are found in vast plutonic bodies in the cores of eroded mountain chains. It is in this environment that the advocates of granitization, that is the production of granite in the solid state by transfusion of alkalis and other elements, find their main lines of evidence. An alternative body of opinion favours the emplacement of such bodies magmatically, the magma having been formed by the remelting of existing crustal rocks within a zone of depressed crust along a zone of mountain-building. This particular controversy seems likely to remain with us for a long time since it deals with the frontier between high-grade metamorphism and igneous activity.

9. Sedimentary rocks

Introduction

The materials derived from the erosion and decomposition of existing rocks are subsequently deposited in rivers, lakes, the sea, or on the surface of the land and are termed sediments. Burial and compaction of sediments, together with the cementing together of constituent grains is the process known as *lithification*, and the resultant product is a sedimentary rock. *Sedimentary petrology*, that is, the study of the mineralogy and texture of sedimentary rocks, will be the main subject of the present chapter.

Unlike igneous rocks, most sediments are made up of a very small selection of minerals, of which quartz, feldspars, the clay minerals, and carbonates such as calcite, are by far the most important. A classification based on mineral assemblages only, would thus prove inadequate to express all the complex features shown by sediments. Most classifications therefore employ textural as well as mineralogical features and as a result, have genetic implications to a far greater degree than most igneous rock classifications. The resulting scheme can be complex, but it does give a convenient means of studying such genetic features as the source area, the mode and extent of transport, and the depositional environment of sedimentary materials.

Before considering a classificatory system, however, it is important to understand the various processes which affect sedimentary materials before and during their deposition. The first part of this chapter deals with these processes, their effects on the mineralogy and chemistry of the parent materials during rock-breakdown, and their effects on the shape, size, and distribution of particles during transport to the depositional area.

Sedimentary Processes and their Effects

Rock Breakdown. All rocks exposed to the atmosphere and ground waters are subjected to varying degrees of physical and chemical weathering. Physical weathering results in the mechanical breakdown of parent materials, producing a residue of mineral grains and rock fragments. In cases of pure physical weathering the bulk chemical composition of the residue is the same as that of the original material. More profound changes take place under conditions of chemical weathering, and the residue may differ markedly in composition from the parental material. The residue in these cases is composed of a *resistate* fraction, and a fraction of newly formed *clay minerals*. The resistate fraction is composed of those constituents of the parental material which have resisted chemical weathering and remain in their original form. Amongst these, quartz is the most abundant and may be accompanied by such materials as rock fragments, feldspar grains, and other rarer constituents such as zircon.

The newly formed clays result from the decomposition of various ferromagnesian and alumino-silicate minerals.

Some of the materials of the original rock are, however, retained neither as resistate nor as clay mineral fractions but are removed as solutions or as colloidal suspensions. The ultimate fate of these elements is varied. Iron and manganese, for example, remain in solution until they encounter an environment which is sufficiently oxidizing for them to be precipitated as oxides. Calcium and phosphorus may be precipitated as carbonates or phosphates in alkaline environments, while much sodium, potassium, and magnesium may remain in solution indefinitely in the oceans. Only extreme evaporation will cause these to be precipitated in any quantity.

Stability of Minerals. During the cooling of igneous magmas, as we have seen on p. 158, minerals tend to appear in a particular sequence which we have referred to as the Bowen reaction series. Even some of the lower temperature minerals of igneous rocks, e.g., orthoclase, are not, however, stable at room temperature. The fact that they are preserved at all is due to the rapid falling off of reaction rates as the rock solidifies and cools. Chemical weathering, involving the presence of water to promote reactions, is concerned with the re-adjustment of minerals which are stable at high temperatures (and sometimes also high pressures) to equilibrium under atmospheric conditions. For this reason the high-temperature minerals of igneous rocks are more prone to weathering than the low-temperature minerals; the latter, having formed at lower temperatures, are already more nearly in equilibrium with the weathering environment. Thus, olivine and pyroxene are very readily decomposed, as are the more calcic plagioclases. Hornblende and sodic plagioclase are somewhat more resistant, potassic feldspars such as microcline more resistant still, while the micas, closely related to the clays, are affected very little, and quartz not at all.

Maturity of Sediments. Complete decomposition of the parental material at the site of weathering or during subsequent transport is rare. Commonly the bulk of the residuum is composed of quartz with subordinate amounts of feldspars and minor amounts of ferro-magnesian silicates. The proportions of these constituents depend, of course, on the nature of the parent material and on the severity of the weathering. During transport, weathering is still effective and the more unstable minerals will become depleted, while quartz increases in proportion. The clays formed by the decomposition of the unstable minerals may be transported with the remaining resistate fraction or may be removed to some other depositional area.

The presence, absence, or variation in the abundance of the more unstable minerals can thus be a guide to the duration of the transport which the sediment has undergone.

Sediments derived from a granodioritic source rock, for example, could be expected, in the immediate vicinity of the source, to contain hornblende, micas, both sodic and potassic feldspars, and quartz. Further away the sediment would be expected to contain potassic feldspar, quartz, and mica, but would be free of hornblende and sodic feldspar. After extensive transport and exposure to weathering, the sediment would have a resistate fraction composed very largely of quartz, with only insignificant amounts of potassic feldspar. The degree to which the mineralogy of the resistate fraction approaches this end-product can be termed the *mineralogical maturity* of the sediment.

Quartz is the only chemically and physically durable mineral common enough to be accumulated in great volume. Hence the mineralogical maturity of the resistate fraction of a sediment can be expressed by its quartz content. Alternatively, since quartz in many igneous and metamorphic rocks is accompanied by feldspar, the mineralogical maturity can be expressed by the ratio of quartz/feldspar. This is a useful general concept, but it must be stressed that detailed comparative studies of the maturity of sediments must always refer to a particular group of sediments derived from a single source area.

Chemical Features of Sedimentary Rocks. Chemical analyses of sedimentary rocks are mainly very different from those of igneous rocks, as a comparison of Table 9.1 with Table 8.1 (p. 151) will show. Analyses of common sedimentary rocks show a much wider range of variation than those of common igneous rocks. Sedimentary processes, for example, are extremely efficient in separating certain elements such as silicon and calcium, with the result that rocks of widely differing compositions are formed.

Table 9.1

CHEMICAL COMPOSITIONS OF SOME TYPICAL SEDIMENTARY ROCKS

	(1)	(2)	(3)	(4)	(5)	(6)
SiO_2	74·14	78·14	99·14	58·10	55·02	7·61
TiO_2	0·15	—	0·03	0·65	1·00	0·14
Al_2O_3	10·17	11·75	0·40	16·40	22·17	1·55
Fe_2O_3 }	0·56	1·23	0·12	4·02	8·00	0·70
FeO }	4·15	—	—	2·45		1·20
MgO	1·43	0·19	Nil	2·44	1·45	2·70
CaO	1·49	0·15	0·29	3·11	0·15	45·44
Na_2O	3·56	2·50	0·01	1·30	0·17	0·15
K_2O	1·36	5·27	0·15	3·24	2·32	0·25
H_2O^+ }	2·66	0·64	0·17	5·00	7·76	0·38
H_2O^- }					2·10	0·30
CO_2	0·14	0·19	—	2·63	—	39·27
C	—	—	—	0·80	—	0·09
	99·81	100·06	100·31	100·14	100·14	99·78

(1) Typical greywacke
(2) Typical arkose. Torridonian Scotland
(3) Devonian orthoquartzite (very pure)
(4) Silty Clay
(5) Residual clay from weathering of gneiss
(6) Allochemical limestone

One particular feature of sediments relative to igneous rocks is their generally high ratio of K_2O/Na_2O. In most igneous rocks Na_2O exceeds K_2O, and in cases where this is not so, e.g., the granites, the excess of K_2O over Na_2O is not marked. Most sediments, in contrast (see Table 9.1), show a considerable excess of K_2O over Na_2O which is caused by the relative durability of potassic feldspars in the resistate fraction, coupled with the taking up of potassium by clay minerals (note, for example, the high potassium content in analyses 4 and 5). Sodium-bearing feldspars decompose early in the weathering process and most of the sodium is carried off in solution, to accumulate in the sea.

A second distinction between sediments and igneous rocks is the generally low calcium content of the former. Incomplete weathering at the source and during transport may cause the retention of some calcium in the resistate fraction, e.g., in hornblende and plagioclase in immature rocks such as greywacke (analysis 1). Normally, however, the early breakdown of calcic plagioclase, hornblende, and augite, ensures the removal of much of the original calcium in solution. The calcium is later precipitated as the carbonate to form limestones (analysis 6) or may contribute to other rocks by the mixing of the precipitate with other constituents, e.g., in the calcium- and CO_2-bearing silty clay of analysis 4.

Silica in sediments is largely held in quartz, while the alumina is concentrated in clay minerals and feldspars. In rocks of low clay content, therefore, the ratio of SiO_2/Al_2O_3 is broadly related to the quartz/feldspar ratio, and may thus be thought of as an index of chemical maturity. Similarly, in rocks free of carbonates, the calcium content gives some indication of maturity. The ratio of Na_2O/K_2O is also useful in this context. It may be observed that the greywacke (analysis 1) shows signs of general immaturity on all three counts.

Textures of Sediments. The effect of transport on the resistate fractions of parent rocks after initial weathering is to change the size and shape of grains. With coarser resistates there is a decrease in size and a decrease in the angularity of the grains with increasing transport. In many cases, abrasion, that is attrition, solution, chipping, and splitting, of grains is the main cause of these changes. Thus turbulent, fast-flowing waters are likely to abrade sedimentary grains faster than calm waters. The decrease in grain size is also related to the resistance of the transported material. Limestone fragments under given conditions are for example abraded more quickly than quartz pebbles.

A second important effect due to the transport of sedimentary materials is the *sorting* of the particles according to their shape, specific gravity, or size. With prolonged transport many sediments undergo size-sorting which tends to produce a relatively uniform grain size. Dune sands, for example, often show a particle size distribution of low standard deviation (due to the efficiency of the wind in separating different sized particles) and may be referred to as *well sorted.* Beach shingle, in contrast, which has undergone little transport, may be composed of pebbles with admixed sand and shell fragments, and as such exhibits a wide range of particle sizes. Such a sediment is said to be *poorly sorted.*

Classification of Sediments

With a knowledge of sedimentary processes we may now turn to the classification of sedimentary rocks. The constituents of sediments may be included under three headings which form the basis of the classification we shall employ. These are:

(a) *Terrigenous constituents*, that is the resistates and clay minerals derived from a source outside the depositional area.

(b) *Allochemical constituents*, that is chemically precipitated materials, formed within the depositional area and showing evidence of subsequent transport. Shell fragments are obvious examples.

(c) *Orthochemical constituents*, materials chemically precipitated within the depositional area and showing no evidence of post-precipitation transport.

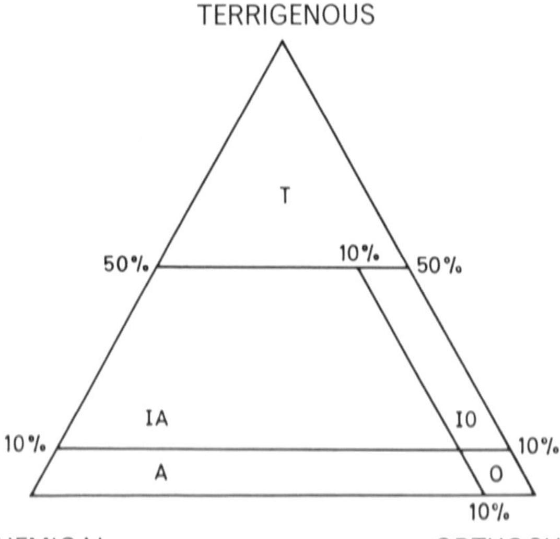

9.1 Classification of sedimentary rocks in terms of terrigenous, allochemical, and orthochemical constituents.

T— terrigenous rocks. A— pure allochemical rocks. O— pure orthochemical rocks. IA— impure allochemical rocks. IO— impure orthochemical rocks.

Few rocks are composed purely of any one of these types of constituent and the basis of the classification is therefore the relative proportions of the three, as shown in the triangular diagram (Fig. 9.1).

Terrigenous Rocks are the most abundant sediments (approximately 75 per cent of all sediments), and include mudstones, siltstones, and sandstones, in decreasing order of abundance.

Allochemical Rocks comprise approximately 15 per cent of sediments and include broken shell beds, oolites (see p. 193), and rocks composed of fragments of limestone derived by penecontemporaneous erosion within the depositional area.

Orthochemical Rocks are relatively rare in their occurrence, comprising approximately 10 per cent of sediments. Very fine-grained limestones belong to this group.

Other groups illustrated in Fig. 9.1 are the *impure allochemical rocks* and the *impure orthochemical rocks*, both of which contain up to 50 per cent of terrigenous material.

THE TERRIGENOUS SEDIMENTS

Mineralogical Composition of Terrigenous Sediments. Unlike igneous and metamorphic rocks, the terrigenous sediments are made up of only a few important minerals, though these may be derived from a great variety of sources. Terrigenous rock fragments on the other hand can display all possible compositions.

The fragments within sedimentary rocks are referred to as *clasts*, and these are often held together by an orthochemical cementing material. The following list is concerned with the mineralogy of the clasts in terrigenous sediments, not with the cement:

Quartz is usually the most abundant clastic constituent of terrigenous sediments and may be derived from a variety of sources. A careful study may give some information about the nature of the source area.

Rock Fragments are mainly of igneous and metamorphic rocks, but also include pre-existing sediments.

Feldspars are abundant in certain sedimentary rocks, the potassic varieties being by far the most important.

Clay Minerals are highly important but cannot normally be studied with the polarizing microscope because of their fine grain. The clay minerals are all closely related, the most important being the potassium-rich *illite* (sericite), aluminium-rich *kaolinite*, magnesium-rich *montmorillonite*, and the iron-magnesium-rich *chlorite*. Not all the clay in a rock is necessarily of terrigenous origin.

Chert, a fine-grained siliceous rock, is mostly derived from previous sedimentary formations.

Coarse Micas include muscovite and biotite, the latter much less abundant.

Carbonates are found as clasts, mainly derived from reworked orthochemical and allochemical limestones.

Accessory minerals commonly include magnetite, ilmenite, hematite, limonite, zircon, tourmaline, and rutile. Relatively unstable minerals such as garnet, apatite, kyanite, epidote, and hornblende, are less common.

Classification of the Terrigenous Rocks. The terrigenous rocks can be sub-divided into three major groups on the basis of grain size as shown in Table 9.2. In poorly sorted rocks there may be mixing of

Table 9.2

CLASSIFICATION OF THE MAIN TYPES OF TERRIGENOUS ROCKS

Sediment	*Rock*	*Textural Term*	*Grain Size*
Gravel	Conglomerate	Rudite or Rudaceous rock	>2 mm
Sand	Sandstone	Arenite or Arenaceous rock	$\frac{1}{16}$–2 mm
Mud or clay	Shale or Mudstone	Argillite or Argillaceous rock	$<\frac{1}{16}$ mm

the three grain-size groups in varying proportions so that compound names such as sandy shale become necessary.

The Conglomerates (*Rudaceous Rocks*). These can be classfied in terms of the size of the dominant rock fragments for which such terms as *boulder* (>256 mm), *cobble* (64–256 mm), and *pebble* (4–64 mm) are commonly employed. Alternatively, nomenclature can refer to the composition of the rock fragments, e.g., granite-sandstone-shale-conglomerate. The shape of the fragments may also be given consideration and rudaceous rocks consisting predominantly of sharply angular rock fragments are distinguished as *breccias*. Other rudites may contain well-rounded fragments of discoid, elliptical, or sub-spherical shapes, while fragments resembling a flat-iron in shape may be found in those conglomerates of glacial origin specifically called *tills*.

The spaces between fragments may be voids but are more commonly filled with finer sediment so that the particle size distribution within the rock is bimodal. In some instances, e.g., beach gravels, the principal mode may fall within the gravel grain size, while in others, e.g., pebbly shales, the principal mode falls within the clay grain size.

Coarse *pyroclastic* rocks, deposits derived from active volcanic sources, form a group apart from normal sediments, although they may be interbedded with and grade into normal terrigenous sediments. The term *agglomerate* is used for pyroclastics having the appearance of conglomerate, while finer grained materials are known as *tuffs* or *ashes*. The sorting is often so extremely poor in pyroclastic rocks that any precise grain size classification is of little value.

The study of rudaceous rocks and the coarser pyroclastics can often be carried out profitably without the aid of a microscope. These rocks are particularly suitable for provenance studies because of the relative ease with which the rock fragments can be assigned to their source area.

The Sandstones (*Arenites*). These can be defined as predominantly siliceous sediments composed mainly of quartz and made up predominantly of clasts within the size range $\frac{1}{16}$–2 mm. In the study of sandstones it is important to look for features which may be of value in provenance studies (i.e., the identification of the source rocks from which the clasts are derived) and in the determination of mineralogical and textural maturity. Sandstones may also give some indication of the density and viscosity of the transporting medium in the ratio of sand grains to interstitial finer grained detrital materials. This aspect of the subject is outside the scope of the present work but is part of the basis of the classificatory system given for sandstones in Table 9.3.

Greywackes and *subgreywackes* are probably the most abundant types of sandstones and are characterized by a considerable (15 per cent) content of mud or silt which acts as a cement after lithification. Unlike some other sandstones, they contain only negligible amounts of orthochemical and allochemical constituents.

Greywackes contain considerable proportions of feldspar and rock fragments, and depending on the relative amounts of these constituents can be designated lithic or feldspathic greywackes.

Microscopic examination of these rocks shows abundant evidence of textural and mineralogical immaturity (see Fig. 9.2). Feldspars such as microcline and orthoclase are usually present and may be

Table 9.3

CLASSIFICATION OF SANDSTONES
(based on Pettijohn)

<table>
<tr><td colspan="2">Matrix, i.e., fine terrigenously derived minerals, e.g., clay minerals</td><td colspan="2">Matrix prominent >15%</td><td colspan="3">Matrix of fine terrigenous Silt or Clay absent or scanty (<15%)</td></tr>
<tr><td rowspan="5">SAND FRACTION</td><td rowspan="2">Feldspar exceeds rock fragments</td><td rowspan="4">GREYWACKES</td><td rowspan="2">Feldspathic greywackes</td><td colspan="2">ARKOSIC SANDSTONES</td><td rowspan="4">ORTHOQUARTZITES</td></tr>
<tr><td>Arkose</td><td>Subarkose or feldspathic sandstone</td></tr>
<tr><td rowspan="2">Rock fragments exceeds feldspar</td><td rowspan="2">Lithic greywacke</td><td colspan="2">LITHIC SANDSTONES</td></tr>
<tr><td>Subgreywacke</td><td>Protoquartzites</td></tr>
<tr><td>Quartz content</td><td colspan="2">Variable; generally <75%</td><td><75%</td><td>>75% <95%</td><td>>95%</td></tr>
</table>

9.2 Typical greywacke, showing poor sorting of angular to rounded clasts. Note the fragment of basic igneous rock in the centre of the section. ×30. Plane polarized light.

accompanied by sodic plagioclase (see the chemical analysis in Table 9.1). Some greywackes of young geological age contain plagioclase as poorly rounded cleavage fragments. Hornblende and augite amongst the ferromagnesian minerals, are not uncommon, though chlorite is by far the most abundant representative of this group and in many cases may have been formed by post-depositional alteration of other minerals.

Lithic fragments frequently consist of chert or fragments of argillites but volcanic fragments such as basalt, andesite, and

9.3 Highly feldspathic and moderately sorted arkose, showing angular to sub-rounded clasts of quartz and microcline. ×30. Crossed nicols.

rhyolite may also be abundant, together with coarser grained igneous rocks. The matrix between the larger clasts is made up of clay minerals, notably finely divided micas and chlorite, some of which may be secondary in origin.

Texturally, greywackes are very variable. Particle shape can vary from well rounded to extremely angular. Sorting is never good and in extreme cases there may be a complete gradation in grain size from the coarser clasts down to the finest matrix.

Arkoses are composed predominantly of feldspars and quartz, and grade with an increase of rock fragments into lithic sandstones, and with a decrease of feldspar into feldspathic sandstones (see Table 9.3). Most arkoses are rather coarse grained, only moderately sorted, and composed of angular to sub-angular clasts (see Fig. 9.3). These features, together with the abundance of feldspar, indicate a fair degree of immaturity. The presence of quantities of feldspar indicates an interruption or retardation of weathering in the source and depositional areas. This may often be due to rapid erosion in the

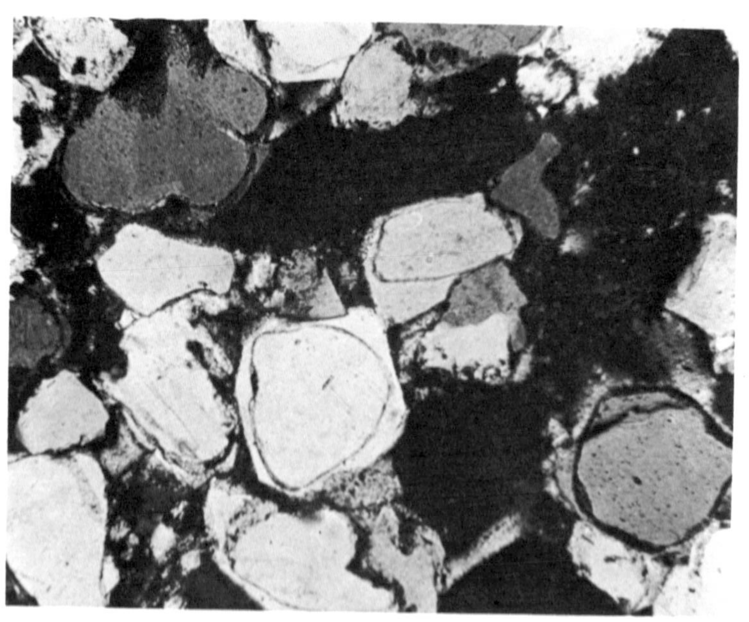

9.4 Rounded to sub-rounded quartz in sandstone containing a carbonate cement. Note the overgrowths in optical continuity with the quartz grains. × 30. Crossed nicols.

source area with resultant rapid burial of sediment and protection from weathering. Small distances of transport are also implied.

The *orthoquartzites* are characterized by a very high quartz content and the virtual absence of other constituents. With an increase of feldspar above the 5 per cent level the orthoquartzites grade into feldspathic sandstones, while with an increase in the content of rock fragments they grade into lithic sandstones, also known as protoquartzites.

A siliceous cement is the most usual, often occurring as overgrowths in optical continuity with quartz clasts. Viewed between crossed nicols such rocks have a mosaic-like texture (see Fig. 9.4) and the original grain boundaries of the clasts are more clearly visible in plane polarized light (see Fig. 9.5). Calcium carbonate and iron oxides may also be found as cementing materials.

Orthoquartzites are usually well sorted and the grains are highly rounded, both features associated with considerable maturity (see

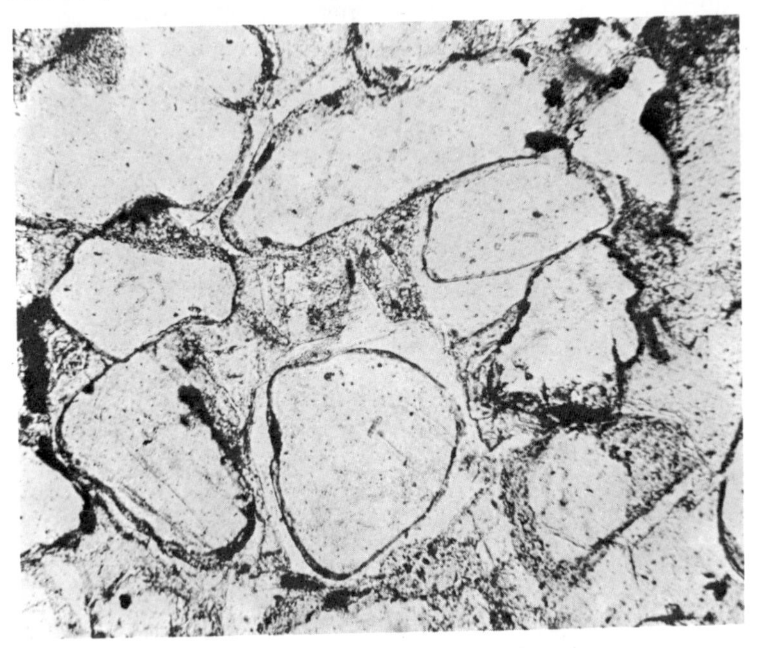

9.5 The same section as in Fig. 9.3 slightly magnified, showing the boundaries of the quartz clasts, overgrowths, and the carbonate cement. ×45. Plane polarized light.

Fig. 9.6). The formation of orthoquartzitic rocks implies a lengthy exposure of the sediment to weathering, sorting, and abrasive processes. Some of these rocks, however, may be formed by the breakdown of existing mature sedimentary rocks and hence their maturity may be thought of as inherited from an earlier sedimentary cycle (see Fig. 9.7).

Cementation of Sandstones. Compaction of sedimentary grains without dissolution is rarély sufficient to produce hard sedimentary rock. Most sediments are subjected to some cementation during lithification.

9.6 Orthoquartzite showing good sorting of sub-rounded to rounded quartz. Feldspar is absent. Careful inspection shows some siliceous cement in grain interstices. ×30. Crossed nicols.

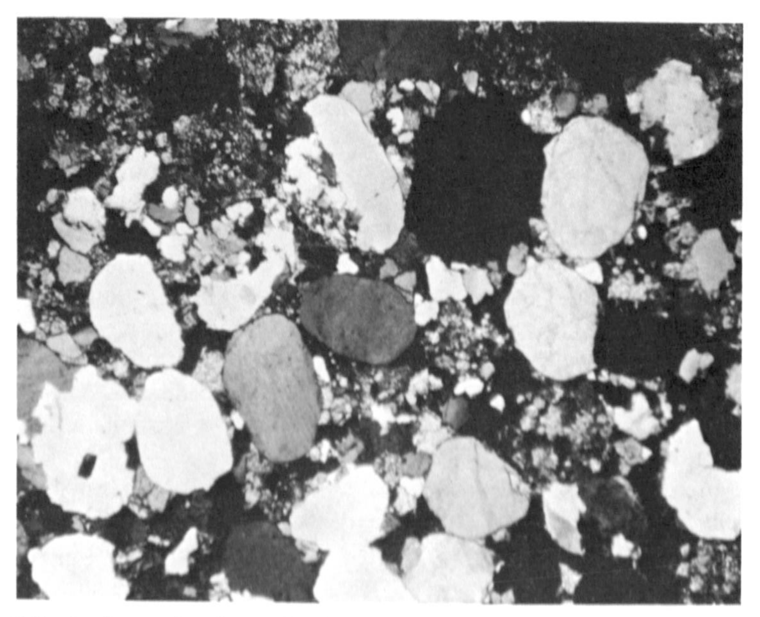

9.7 Sandstone showing a bimodal size distribution of large rounded quartz and smaller angular quartz set in carbonate matrix. ×30. Crossed nicols.

Commonly sediments subjected to the weight of an overburden show strain at points where adjacent clasts press upon each other. This may result in dissolution of the clasts at these points and reprecipitation of the silica in the interstitial spaces. Cementing materials may also be produced by chemical precipitation during clastic sedimentation, as is the case with some carbonate and ferruginous cements. More often, however, carbonate cements result from the recrystallization of organic debris and fine precipitated carbonates, which are recrystallized either in situ or elsewhere within the adjacent rock. Ferruginous cements are often found in sediments which have been deposited within environments of oxidizing conditions, e.g., flood-plain and terrestrial environments, where iron-bearing solutions derived from the dissolution of minerals may subsequently reprecipitate their iron as hematite or limonite when conditions of oxidation are met.

The Argillaceous Rocks. Argillaceous rocks are the most abundant of all sediments and consist predominantly of clays (particle size range less than 0·004 mm) and silts (0·004–0·060 mm), the latter being mainly composed of chips of quartz, feldspars, and mica. Because of the fine-grained nature of the argillaceous rocks, studies using the microscope are rarely satisfactory, and recourse must be had to various techniques such as X-ray diffraction and infra-red and differential thermal analyses, in order to identify the clay minerals present. At best, it may be possible to identify particles within the silt range using the microscope.

Argillaceous sediments may contain considerable quantities of carbonaceous matter and may approach black in colour as a result. The carbonaceous matter forms brownish and blackish streaks and pods, sometimes defining the bedding. More often, however, bedding in shales is defined by alternations of layers of differing grain size.

Many carbonaceous shales contain euhedral grains of pyrite formed as a result of the post-depositional reconstitution of disseminated sulphur or iron sulphide within the sediment. Micro-organisms probably play an important part in the precipitation of pyrite.

In contrast with carbonaceous shales, which are deposited in reducing environments, red and red-brown shales are formed in oxidizing environments and owe their colour to the presence of iron oxides such as hematite and limonite.

Argillaceous rocks may also contain significant amounts of carbonates and are then known by the general term of *marl*. Carbonates such as calcite, dolomite, and siderite, together with phosphates and sulphates such as barite, may form nodules and concretions within shales.

General Scheme for the Description of Terrigenous Rocks. It is important to discuss features of mineralogy and texture in some logical order and the following scheme is intended to give some guidance in describing samples of terrigenous sediments. For a particular rock, however, not all the headings given below will necessarily be appropriate.

(a) Name of Rock.

(b) Megascopic properties, e.g., hardness, colour, bedding, lamination, fissility, fossils, sedimentary structures such as ripple-marks.

(c) Microscopic Description.

(i) Brief summary covering prominent types of clasts, shape and size of clasts, nature of cementing material.

(ii) Mineralogical composition of clasts in whatever detail may be necessary. This will include quartz, rock fragments, feldspars, clay minerals, terrigenous chert, coarse micas, accessory minerals, e.g., iron ores and zircon, allochemical and orthochemical minerals, e.g., intraclasts (see p. 193), ooliths (see p. 193), microfossils.

(iii) Estimate the percentage of the various terrigenous constituents and derive a mineralogical maturity index from the quartz/feldspar ratio. Also for purposes of nomenclature determine the ratios of feldspar/rock fragments and terrigenous mud to coarse and medium clasts. Estimate the relative proportions of terrigenous, orthochemical, and allochemical constituents.

(iv) Texture. Degree of sorting of terrigenous constituents. Terms such as well sorted, moderately sorted, and poorly sorted may be used.

Shape of terrigenous clasts. It may be necessary to consider clasts of different materials separately. Terms such as well rounded, moderately well rounded, subangular, angular, extremely angular, may be used.

An assessment of textural maturity should be made, based on the observations above. Anomalies should be noted, e.g., poor sorting coupled with good rounding.

(v) Cementation. Type of cement, whether in optical continuity with clasts, etc.

(vi) Evidence of recrystallization. Overgrowths, solution effects between adjacent grains.

(d) Inferences. Brief description of the possible mode of origin of the rock, including the source area, the nature of the weathering and transport and depositional environments, the post-depositional history.

THE ALLOCHEMICAL AND ORTHOCHEMICAL ROCKS

Carbonate rocks, or more specifically limestones, are the most important rocks of the group. Limestones typify the conditions met with in chemical sedimentation and will be discussed more fully than the other chemical sediments.

The Limestones. Limestones, including dolomite (magnesium-rich carbonate rocks) comprise some 25 per cent of all sedimentary deposits. The term limestone is normally used for rocks containing 50 per cent or more of calcium carbonate. They are widely variable in appearance and in mode of formation. Because of the complexity of post-depositional alteration which these rocks undergo, they are generally more difficult to study than the terrigenous sediments.

In general, in limestones which are not in an advanced state of recrystallization it is possible to recognize both orthochemical and allochemical constituents. The *orthochemical* constituents are formed by normal processes of chemical precipitation within the basin of deposition and show little evidence of post-depositional transport. These constituents include *microcrystalline calcitic ooze* having a grain size in the range 0·001–0·005 mm (similar to that of the clay minerals) which in general cannot be satisfactorily resolved under the microscope. This sort of calcite mud results from normal inorganic or biological precipitation of calcite in areas of weak bottom currents. Also included under the orthochemical constituents is *sparry calcite cement* (grain size greater than 0·01 mm) consisting of clear, resolvable grains of calcite. It has been shown that sparry calcite is seldom mixed in large amounts with microcrystalline ooze.

The *allochemical* constituents (often referred to simply as allochems) show evidence of transport after precipitation of the original carbonate material and include:

(a) *Fossils and shell fragments.* These may be abundant enough to justify such names as foraminiferal limestone, coral limestone, etc.

(b) *Ooliths.* These are spherical or subspherical carbonate aggregates having a radial or concentric structure (see Fig. 9.8). They usually exceed 0·1 mm in diameter and are often centred on nuclei such as sand grains, shell fragments, etc. The formation of oolites usually results from the vigorous and continuous wave action in areas where calcite is being precipitated.

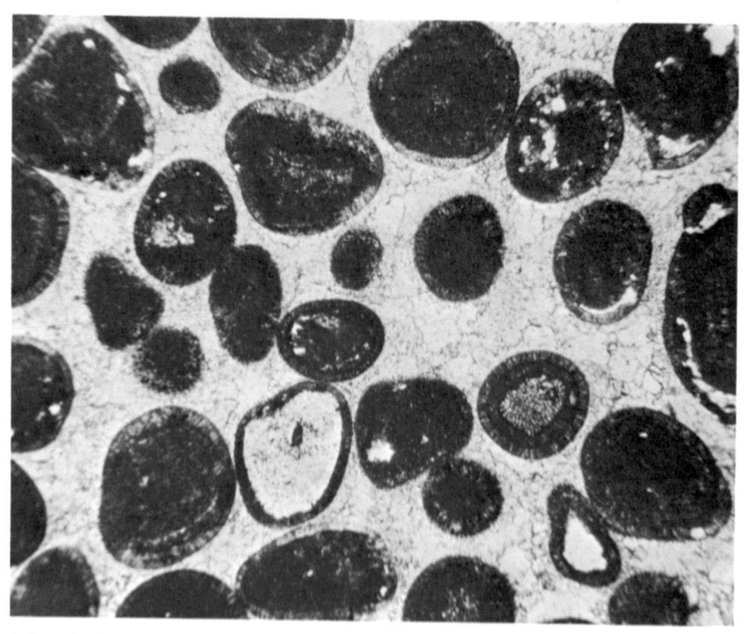

9.8 Typical oolite showing predominantly concentric ooliths set in sparry calcite matrix. × 30. Plane polarized light.

(c) *Intraclasts* are fragments of consolidated or slightly consolidated limestone derived from within the basin of deposition. They are genetically distinct from terrigenously derived fragments of pre-existing limestone. Intraclasts result from local erosion in parts of the basin of deposition during the depositional period. They may be recognized as clasts within a limestone by their distinct outline in thin sections.

Classification of Limestones. Limestones may be classified in a manner similar to that employed for the terrigenous rocks, on the basis of the grain size of the allochemical constituents. Thus, three groups may be distinguished:

(a) Calcirudites—fragments greater than 1 mm.
(b) Calciarenites—fragments 0·06–1·0 mm.
(c) Calcilutites—grain size less than 0·06 mm.

Some confusion exists over limestone nomenclature since some texts use the terms calcirudite and calciarenite for conglomerates and sandstones formed of reworked clasts of pre-existing terrigenous rocks. To avoid confusion it is perhaps best to prefix the rock name by the term allochemical when the classification is based on allochemical grain size as it is here.

It is possible to classify limestones further by a consideration of the proportions of three constituents, i.e., allochems, sparry calcite cement, and microcrystalline ooze. The proportions have some genetic significance since a high proportion of ooze indicates calm conditions of precipitation and will in general coincide with the absence of ooliths. Absence of ooze may signify more turbulent conditions while allochems such as large intraclasts indicate highly turbulent conditions.

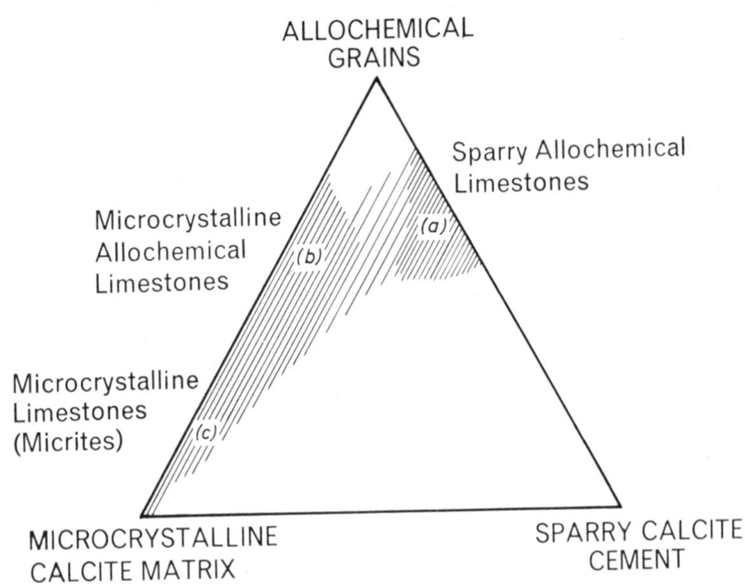

9.9 The three main types of limestone. (a) sparry allochemical limestones; (b) microcrystalline allochemical limestones; (c) microcrystalline limestones (micrites). Shaded areas represent main fields of limestones when plotted in terms of the three end members.

A triangular plot of limestones in terms of the three constituents is given in Fig. 9.9 and shows that the great majority of the rocks fall into two distinct areas, one of which may be sub-divided. Thus limestones may be considered under three main headings:

(a) *Sparry allochemical limestones.* These consist of allochemical constituents bound by a calcite cement (see Fig. 9.10). Many fossiliferous and oolitic limestones are of this type. The presence of

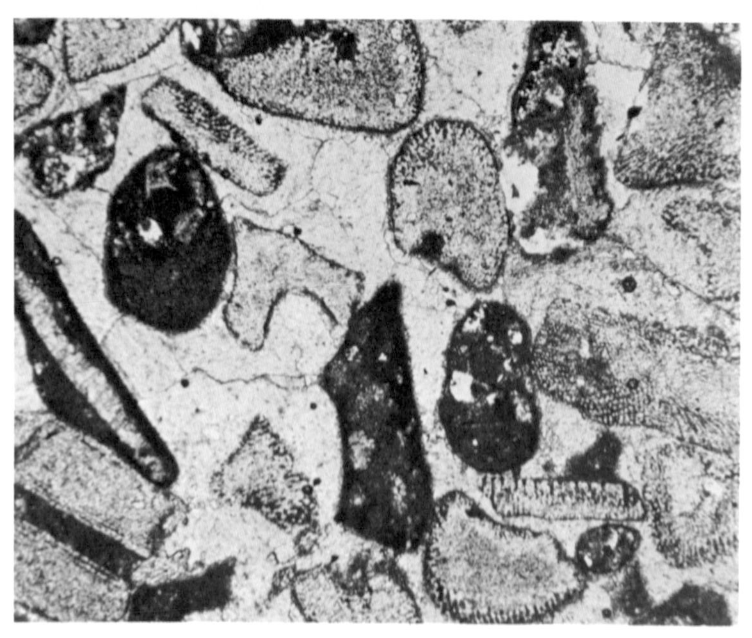

9.10 Sparry allochemical limestone, composed primarily of echinoid debris set in a sparry calcite cement. Grain boundaries of the sparry calcite can be faintly observed. $\times$ 30. Plane polarized light.

aggregated allochemical constituents and the absence of micro-crystalline ooze suggest that currents were sufficiently strong to remove ooze but insufficient to remove coarse sparry calcite, which remains packed between the allochemical clasts. Varying sizes of clasts and variation in the proportions of allochems to sparry calcite produces a wide range of textures. Often the size and shape of the allochems and the packing of the cement is insufficient to produce a homogeneous mass, and the resulting limestone has a high porosity.

(b) *Microcrystalline allochemical limestones* may have the same allochems as the preceding type except that oolites are generally absent. The cement in these types is microcrystalline ooze suggestive of deposition in quiet waters (see Fig. 9.11), or alternatively of rapid burial. Some indication of current strengths may be gained by a study of any admixed terrigenous constituents. The microcrystalline limestones in general have a much lower porosity than the sparry limestones.

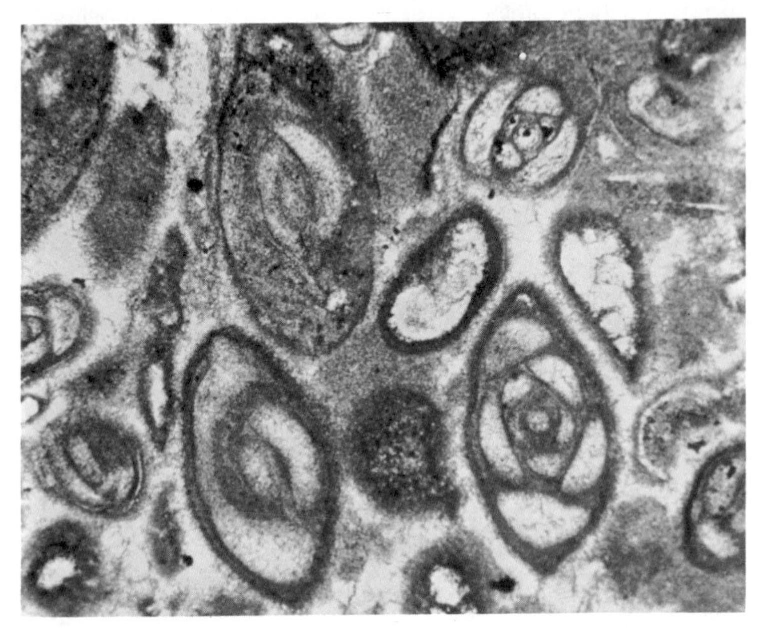

9.11 Microcrystalline allochemical limestone, showing foraminifera set in a microcrystalline ooze (dark grey). Post-depositional alteration has replaced some of the ooze by distinct calcite crystals (clear patches). ×30. Plane polarized light.

(c) *Microcrystalline oozes*, also called *micrites*, are formed in depositional environments where the rate of deposition of allochems is low. Their fine grain size, like that of their terrigenous equivalent shale, implies a low porosity after consolidation.

Post-depositional Alteration of Limestones. More than any rock, limestone is prone to alteration, which may be intense enough to obscure the original nature. Classificatory difficulties may be encountered when microcrystalline cement is recrystallized to sparry calcite. Evidence of recrystallization should always be sought when examining limestones and may be shown by:

(a) A clotted appearance in hand-specimen caused by a patchy recrystallization (particularly in fine-grained rocks) shown in thin sections by areas of coarsely crystalline, clear carbonate grains (see Fig. 9.11).

(b) Replacement of carbonate grains by quartz or chert. Pseudomorphs retaining the original shape of the carbonate grains may be found.

(c) The presence of *dolomite*, often formed by pre- or post-consolidatory replacement of calcite. Advanced optical techniques are required for the identification of dolomite as opposed to calcite, but the distinction can readily be made by the use of chemical staining procedures.

Other Chemically Deposited Rocks. *Chert* is a rock composed of microcrystalline quartz and chalcedony and is commonly found in limestones and dolomites as irregular concretions, and replacement patches and veins. It is also found in terrigenous sediments such as shale.

In thin sections the quartz in the chert is often microcrystalline and is usually contaminated with clay, carbonates and/or iron oxides;

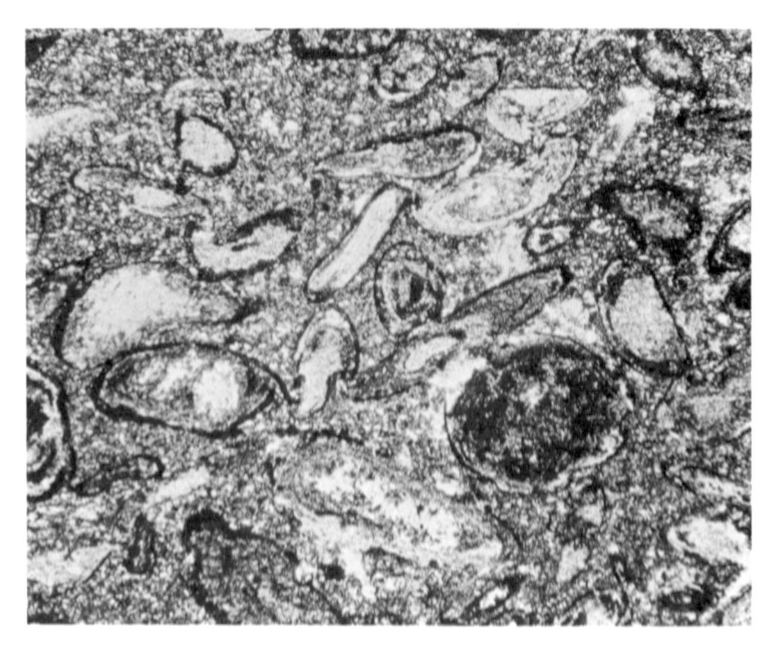

9.12 Chamositic iron ore containing ooliths and spastoliths. The dark bands in the ooliths are probably composed of limonite, while the clear areas are dominantly carbonates. × 30. Plane polarized light.

sometimes it contains well-preserved fossil remains, as is the case with radiolarian chert. Sponge spicules and algal structures are also commonly preserved while concentric siliceous oolites are also found.

It appears that cherts can be both primary and secondary in origin. Many bedded cherts may be primary deposits formed by the agency of micro-organisms. Secondary cherts may predate or postdate the solidification of the rocks in which they are found. The common chert concretions of limestones are mainly of the latter variety.

Many *iron-rich sediments* are of economic importance as iron ores and contain their iron either as siderite (iron carbonate), chamosite (iron silicate), or as the oxides, hematite and limonite. Chamositic iron ores frequently contain ooliths and spastoliths (crushed ooliths), of chamosite (see Fig. 9.12).

The rocks known as *evaporites* are chemically deposited when conditions are favourable for the complete or near-complete evaporation of sea water or other saline waters. The deposits are of great commercial importance as a source of rock salt, gypsum, anhydrite, and a great variety of rarer minerals including, particularly, potassium compounds. Evaporite minerals are prone to recrystallization and alteration.

10. Metamorphic rocks

Introduction

Rocks buried below the surface of the earth are sometimes subjected to temperatures, stresses, and chemical conditions other than those operating during the lithification of sediments. Providing that these physical and chemical conditions are not such as to induce large-scale melting, the rocks respond by undergoing alteration while remaining essentially solid. This process of transformation in the solid state is termed metamorphism, and its products are the metamorphic rocks.

Depending on the extent to which the metamorphic conditions imposed upon a rock differ from the conditions of its formation, metamorphism will involve some *recrystallization* of existing minerals, the formation of new minerals (*neomineralization*) with the elimination of some of the existing ones, and the formation of new textures and structures. By these means the rock strives to achieve mineralogical and structural equilibrium under the new conditions, and the extent to which it succeeds depends upon how long the conditions are operative and the rate at which the various changes proceed.

Many metamorphic changes are of a virtually *isochemical* nature, that is to say there is little change of the chemical composition of the rock during metamorphism. Two important constituents, water and carbon dioxide, are, however, particularly mobile and are commonly added to, or removed from rocks in significant amounts during metamorphic changes. Thus, if we compare sedimentary limestones with their metamorphic equivalents, it is often apparent that CO_2 has been driven from the rock during metamorphism. Similarly metamorphosed argillaceous sediments show a loss of water when compared with their unmetamorphosed equivalents. Conversely, many basic igneous rocks are originally almost devoid of hydrous phases such as amphiboles but are frequently found as amphibole-rich metamorphic rocks. Evidently in these cases there has been a significant addition of water to the system.

It is important to realize that most of the minerals in the metamorphic rocks we see have been formed under conditions consequent upon a rise of temperature (*prograde metamorphism*), and that mineralogical changes following the decline of temperature (*retrograde metamorphism*) are often negligible. This bias in nature towards prograde rather than retrograde changes is partly a result of such features as low reaction rates under conditions of declining temperatures, but is largely a result of a lack of available H_2O and CO_2. Thus, the volatile constituents, H_2O and CO_2, are expelled from the rocks with rising temperature and cease to be available for the formation of low-temperature minerals (usually rich in H_2O and CO_2) when the temperatures decline at the end of prograde metamorphism. In the examination of metamorphic rocks we are, therefore, largely concerned with mineral assemblages formed during prograde metamorphism.

Under many conditions of prograde metamorphism H_2O and CO_2 appear to have been present in sufficient quantities to allow the free formation of those hydrous and CO_2-bearing minerals stable under the given conditions of temperature and hydrostatic pressure (see below). Metamorphic rocks in which this condition does not appear to have been fulfilled will not be considered in this chapter. For reasons of simplicity, we shall also exclude metamorphic changes involving the process known as *metasomatism*, in which the large scale migration of chemical constituents other than H_2O and CO_2 is involved. For the rocks we shall be considering, therefore, the chemical conditions of metamorphism are largely defined by the bulk chemical composition of the pre-existing rock.

The Agents of Metamorphism

The most important physical conditions which cause and control metamorphic processes are temperature and stress. The stress within a body of rock may always be resolved into two components, one of which is referred to as hydrostatic pressure and the other as shearing stress. The former is that part of the stress which may be considered as acting uniformly in every direction, while the latter is the remaining 'unbalanced' stress. The effects on metamorphic processes of the two stress components differ greatly and we shall thus consider the nature of metamorphism to be controlled by three physical variables, namely: *temperature*, *pressure* (meaning hydrostatic pressure), and *shearing stress*.

Temperature exerts a powerful influence on metamorphism because of its direct control of the stability ranges of individual minerals and mineral assemblages. In addition it influences the rate of such processes as diffusion and hence the time taken for a rock to reach equilibrium under new conditions. Pressure, like temperature, directly affects mineral stability ranges and is largely a function of the depth of burial of the rock, though it may be supplemented by stresses of tectonic origin.

Compared with temperature and pressure, shearing stress has virtually no direct influence on mineral stability, and its principal importance lies in its being the casual agent of such processes as *deformation* (mechanically induced changes of shape in rocks). Because of this, shearing stress exerts a profound influence on the textures and structures of metamorphic rocks. However, it does have an indirect effect upon mineralogy, in that deformation of minerals may promote reaction rates, resulting in neomineralization and the elimination of metastable phases, and recrystallization.

Field Classification of Kinds of Metamorphism

Consideration of the distribution of metamorphic rocks leads to the recognition of three principal kinds of metamorphism on the basis of field criteria. These are:

(a) *Dislocation metamorphism*—that occurring along zones of intense deformation such as faults.

(b) *Contact metamorphism*—that occurring in restricted areas adjacent to bodies of igneous rock. The resultant metamorphic rocks form an *aureole* around the igneous body.

(c) *Regional metamorphism*—that occurring over very extensive areas (often thousands of sq. km). In contrast to dislocation and contact metamorphism, this kind of metamorphism is not clearly related to any local event or events. It is particularly characteristic of orogenic or mobile belts.

Although the above definitions involve no mention of the three physical controls or agents of metamorphism, delineated in the preceding section, we find (as might be expected from the geological situations referred to above) that certain combinations of these agents are more common in one kind of metamorphism than in another. The principal agent of dislocation metamorphism is always shearing stress, while temperatures during such metamorphism are usually low, though there may be some generation of heat as a result of friction. Consequently the chief effects of dislocation metamorphism are seen in the textures and structures of the resultant rocks, the amount of recrystallization and neomineralization being highly limited. In contrast, areas of contact and regional metamorphism usually show rocks which have been reconstituted over a large range of temperatures (see the following section on *grade*), and include many rocks formed under high metamorphic temperatures. Considerable recrystallization and neomineralization are therefore associated with contact and regional metamorphism.

In contact aureoles the increase of temperature, consequent upon magma intrusion, is the dominant agent of metamorphism, while pressure is usually low and shearing stress negligible. The products of contact metamorphism therefore differ radically from those of dislocation metamorphism, not only in the extent to which mineral growth has occurred, but also in textures and structures. Metamorphic rocks resulting from regional metamorphism fall between the two extremes represented by the products of dislocation and contact metamorphism, for increased temperature is usually combined with high shearing stress, and therefore, mineral growth takes place in association with deformation. The magnitude of pressure during regional metamorphism may vary widely.

CONTACT AND REGIONAL METAMORPHISM

Grade. During contact metamorphism there is a general decline of temperature with distance from the igneous body responsible for the metamorphism. As a result, providing sufficient time is allowed for the rocks to adjust to this variation in temperature, the metamorphic products will vary in character in different parts of the aureole. Similarly, examination of regionally metamorphosed rocks shows that some have formed under conditions of low temperature and others under conditions of high temperature. In order to relate such variations in the conditions of metamorphism, either within a single metamorphic area or between separate metamorphic areas, the term *grade* is used. Thus, metamorphism occurring at low temperatures is described as *low-grade*, while metamorphism occurring at high temperatures is described as *high-grade*. Differences of grade are shown by such features as a generally coarser grain size in higher grade rocks, and more especially by changes of mineral assemblages in rocks of given chemical composition. Although the detailed study of such changes is beyond the scope of the present work, we shall nevertheless find the concept of grade useful in the discussion of contact and regionally metamorphosed rocks.

Textures and Structures. These may be divided into two broad groups which are considered in turn below.

Textures and Structures of a purely Metamorphic Origin. Much variation in the structure of metamorphic rocks is dependent on the extent to which mineral growth has been combined with deformation, and it will be pointed out in the following pages that many structures are more typical of regional than contact metamorphism and vice versa. Compared with igneous and sedimentary rocks, however, the structures and textures of both contact and regionally metamorphosed rocks show many common and distinctive features resulting from their reconstitution in the solid state. To those textures and structures of metamorphic rocks resulting from the growth of minerals in a solid medium the adjective *crystalloblastic* is applied. Comparing the textures and structures of metamorphic rocks with those of igneous rocks (where growth of minerals has occurred in a liquid) we may note the following distinctive features of those of a crystalloblastic nature:

(a) A much larger proportion of the minerals in metamorphic rocks show little or no crystal form. This is particularly marked for such minerals as the feldspars which are almost invariably anhedral (or xenoblastic—see below) in metamorphic rocks.

(b) Minerals which commonly occur in crystals of inequidimensional shape (e.g., the platy habit of micas; the prismatic or acicular habits of amphiboles) often have their characteristic habits exaggerated in metamorphic rocks as compared with igneous rocks. Furthermore, according to the magnitude of shearing stress during metamorphism, such inequidimensional grains tend to lie in definite orientations in metamorphic rocks giving rise to such structures as schistosity (see below)*.

(c) The larger crystals in metamorphic rocks frequently contain inclusions of small mineral grains. Also, these inclusions are

* It may also be noted that equidimensional minerals such as quartz and calcite may also take up definite crystallographic orientations in response to shearing stress during metamorphism. In these cases, however, the preferred orientation may only be detected by using advanced techniques beyond the scope of this work.

often arranged in some regular manner which may be:

(i) A result of the preservation, during the growth of the larger crystal, of some arrangement possessed by the smaller mineral grains before they become included in the larger crystal.

(ii) A result of the mechanics of growth of the larger mineral. Thus the mineral andalusite sometimes constrains its inclusions to lie only along certain crystallographic planes. Consequently a cross-section of an andalusite prism of this variety (chiastolite) shows the inclusions concentrated along the arms of a cross (see Fig. 7.27).

(d) Compositional zoning and twinning of mineral grains, which is particularly well shown by, for example, the feldspars in igneous rocks, is much less frequently seen in metamorphic rocks.

In the following pages of this section some of the terms used to describe the commoner textural and structural features of contact and regionally metamorphosed rocks are listed, and some discussion of various textures and structures thus defined is given.

*Idioblastic**—describes the shape of a mineral (*idioblast*) showing well-developed crystal form.

Xenoblastic—describes the shape of a mineral (xenoblast) showing no regular crystal form.

The terms *euhedral*, *subhedral*, and *anhedral*, which are used for the minerals in igneous rocks (see p. 161), may also be used to describe the shapes of minerals in metamorphic rocks. Idioblastic and xenoblastic are, therefore, approximately equivalent to euhedral and anhedral respectively.

Granoblastic—this term is applied to those textures in which the mineral grains are all approximately equidimensional and of approximately equal size (see Fig. 10.1).

Porphyroblastic—a texture where grains of one or several minerals are developed to a conspicuously larger size than grains of the other minerals in the rock. The larger mineral grains are termed *porphyroblasts*, and the smaller mineral grains are often said to constitute the *matrix* of the rock.

* The syllable '*blast*' appears in many of the textural and structural terms employed for metamorphic rocks. When used as a suffix, as above, it indicates the texture or structure to be of metamorphic origin. When used as a prefix (see later section on relict structures) it indicates the texture or structure to be a relic surviving from the original rock.

Poikiloblastic (*sieve-structure*)—a structure where some mineral grains (*poikiloblasts*) contain numerous smaller crystals of other minerals. Poikiloblasts are, therefore, usually also porphyroblasts.

Decussate—that structure in which inequidimensional minerals (usually micas and amphiboles) are orientated in all directions, and thus show a 'criss-cross' arrangement in thin section. Decussate structure only forms where shearing stress has been negligible as an agent of metamorphism, and it is only seen, therefore, in the rocks

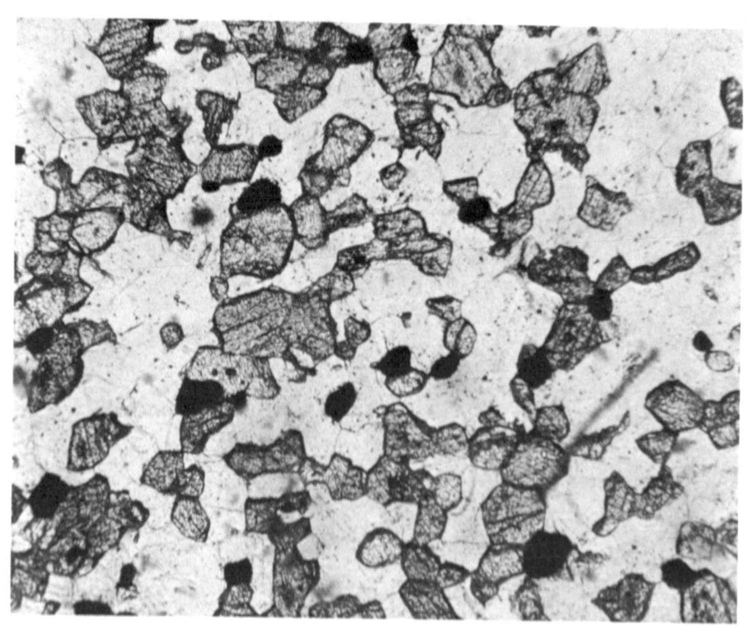

10.1 Granoblastic texture in a high-grade metabasalt (consisting of plagioclase, pyroxene, and iron ore). × 30. Plane polarized light.

of contact aureoles. However, where a marked parallel arrangement of minerals was present in the rocks before metamorphism, decussate structure may not be developed even under conditions of negligible shearing stress, as a result of the metamorphic minerals mimicking the pre-existing mineral arrangement.

Schistosity—that structure formed by the occurrence of inequidimensional minerals in approximately parallel orientation (see Figs. 10.2 and 10.3). Where schistosity is a result of the parallel orientation of platy or tabular minerals (e.g., micas) it is always of a planar kind, but where it is a result of the orientation of elongated (prismatic, acicular, or fibrous) minerals (e.g., amphiboles) it may have either a planar or linear nature. In the case of a linear schistosity the elongated mineral grains lie with their long axes parallel to one another, whereas such minerals form a planar schistosity when their long axes all lie within one plane but show no preferred orientation within that plane. A well-developed planar schistosity when present in fine-grained metamorphosed argillaceous sediments is often referred to as *slaty cleavage.* Rocks possessing schistosity are

10.2 Schistosity formed by the parallel orientation of muscovite and biotite flakes in a quartz-mica-schist. Also showing foliation as a result of the aggregation of the mica into distinct lenticles. $\times$30. Crossed nicols.

described as *schistose*, and they usually split easily along a certain direction or directions. Since schistosity only develops when shearing stress is an important agent of metamorphism schistose rocks are typically found in areas of regional metamorphism.

Metamorphic rocks are often described as showing *cleavage.* When the latter term is used in connection with rocks (as distinct from minerals) it indicates the presence of some structure (of metamorphic origin) which results in a tendency for the rock to split along certain closely spaced approximately parallel surfaces. Planar schistosity (including slaty cleavage) exemplifies a type of rock cleavage known as a *true cleavage*, a name that implies that the rock will split parallel to the cleavage direction at virtually any point chosen within the rock. True cleavage is to be distinguished from rock cleavage of a type known as *false cleavage*, where a rock will only split parallel to the cleavage direction at a limited number of points. A false cleavage may be of several kinds depending upon its cause. Thus in some cases (*fracture cleavage*) a false cleavage results

10.3 Schistosity formed by the parallel orientation of hornblende prisms in a plagioclase-hornblende-schist. ×30. Plane polarized light.

from the fracturing of a rock along a set of closely spaced parallel surfaces, which may be either miniature joint or fault planes. In other cases (*strain-slip cleavage*) a false cleavage is the product of very small-scale folding of a rock possessing true cleavage.

Within an area of regional metamorphism it is common to see several rock cleavages, which are often not only of different types but of the same type in various orientations. Sometimes, especially in low-grade rocks, several cleavages (e.g., one true cleavage associated with one or more false cleavages) may be seen in a single specimen (see Figs. 10.4 and 10.5). Each of such cleavages has usually formed in a separate episode of deformation, but often the deformations have been closely associated in time. The multiplicity of cleavages reveals the complex series of events, characterized by repeated phases of deformation, which occur during one period of mountain-building or orogeny.

Foliation—that structure resulting from the aggregation (during metamorphism) of particular minerals of the metamorphosed rock into layers, lenticles or streaks.* Foliation is commonly associated

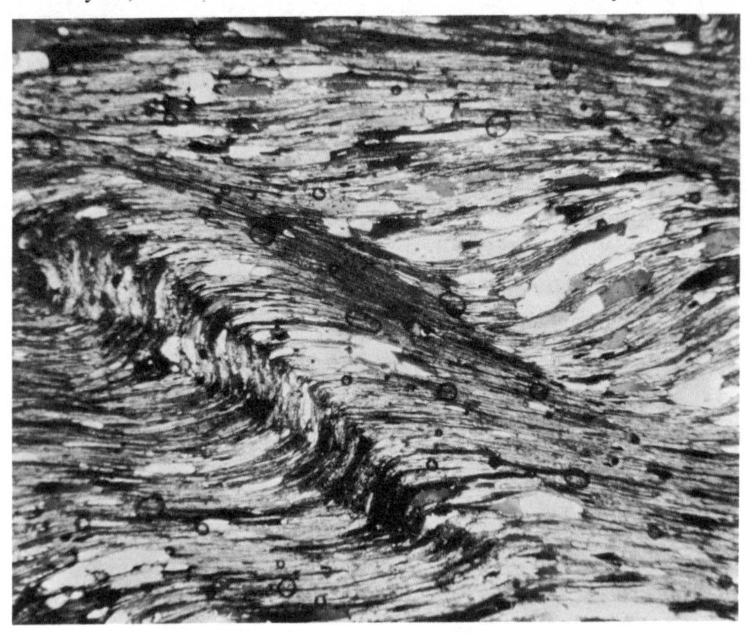

10.4 Pelite with a principal cleavage direction resulting from a mica schistosity, but also showing a strain-slip cleavage formed by small scale folding of the schistosity. ×60. Crossed nicols.

with schistosity, and the two structures, when present in the same rock, are parallel. For example, in a regionally metamorphosed rock consisting of muscovite, biotite, quartz, and plagioclase it is usual to see a schistosity formed by the parallelism of mica plates, parallel to which is a foliation showing the micas concentrated into separate lenticles from the quartz and plagioclase (see Fig. 10.2).

* The usage given here is that favoured by British geologists. American geologists most frequently use 'foliation' as a general name for any type of planar structure of metamorphic origin (thus foliation in this sense would include schistosity for example).

Granulose—this adjective is used to describe the structure of a rock in which most of the minerals are of equidimensional habit, and in which any inequidimensional minerals present do not occur in parallel orientation. The structure is, therefore, commonly seen in rocks consisting largely of such minerals as quartz, feldspar, calcite, and pyroxene, irrespective of whether the rocks have formed in contact or regional metamorphism.

Gneissose—describes the structure of coarse-grained, usually

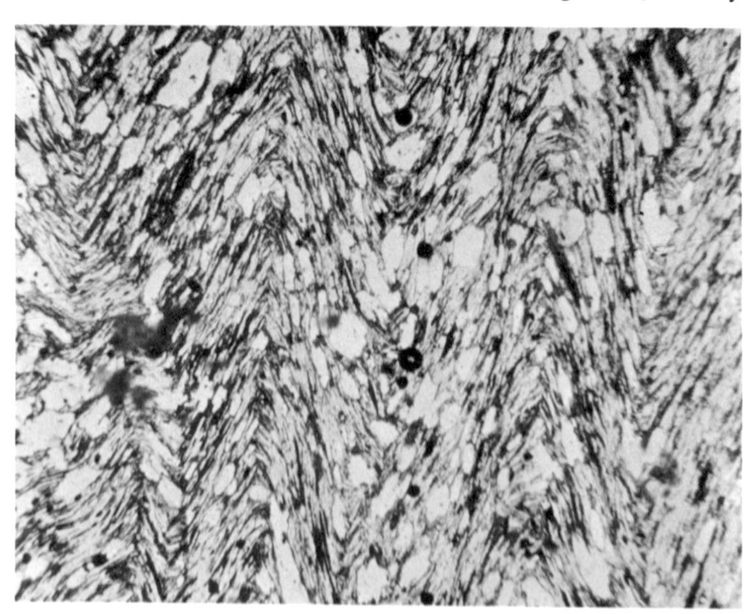

10.5 Pelite whose principal cleavage direction is defined by a strain-slip cleavage, the latter resulting from the extensive small-scale folding of a mica schistosity. ×60. Plane polarized light.

high-grade, rocks with a distinctly banded distribution of minerals. The banding is often the result of the development of a very coarse foliation, but is also often dependent upon some original banding (such as bedding) present in the rock before metamorphism. Frequently the bands in a gneissose rock are of alternately granulose and schistose structure.

Maculose or spotted—these adjectives simply describe the structure of a rock with a distinctly spotted appearance in hand-specimen. The spotting may result from local recrystallization or neomineralization, or from the presence of porphyroblasts whose presence is not otherwise clear in hand-specimen. Spotted structure is frequently seen in low- and middle-grade contact metamorphosed rocks, but may also be found in low- to middle-grade regionally metamorphosed rocks where the regional metamorphism has been of a low-pressure type (e.g., Abukuma type—see p. 226).

Relict Structures. Because the metamorphic alteration of a rock proceeds essentially in the solid state, it is not unusual for some of the structures of a pre-existing rock to remain preserved within the

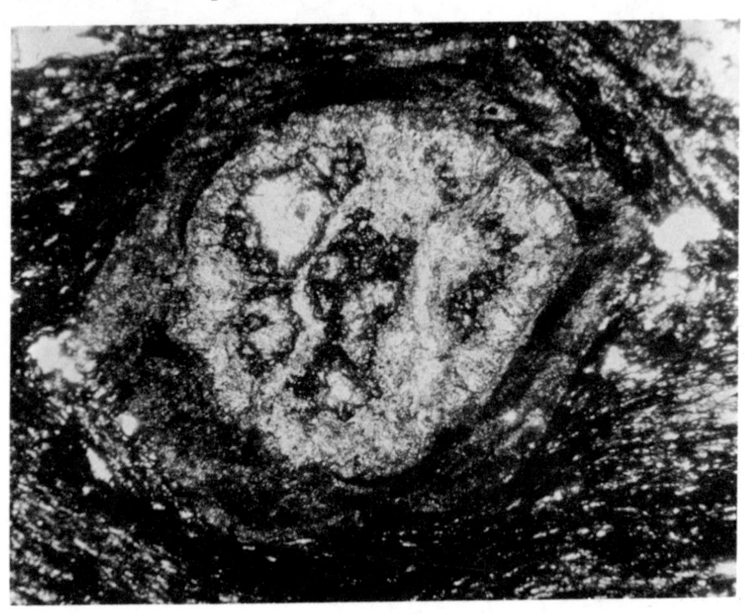

10.6 Pseudomorph after a garnet in a graphitic schist. The garnet originally had well-developed crystal form but is now largely replaced by chlorite showing a dark outer zone and a lighter inner zone. Some remnants of the original garnet are preserved in the inner zone. ×30. Plane polarized light.

metamorphic rock formed from it. The bedding of sedimentary rocks often survives metamorphism, though it may be modified or lost where strong deformation has occurred. In igneous rocks, relics of such features as ophitic or porphyritic texture and amygdaloidal structure may be preserved, even though the original minerals have been replaced by metamorphic minerals. Thus the shape of an original phenocryst of augite, for example, may be identifiable even though the augite has been replaced during metamorphism by an aggregate of hornblende crystals. In cases such as this the later developed minerals are said to *pseudomorph* the original mineral. Metamorphic rocks showing relics of original porphyritic or ophitic texture are described as *blastoporphyritic* and *blastophitic* respectively. They are particularly common in low-grade contact metamorphism, where neomineralization and recrystallization have been limited and deformation usually negligible.

Where, consequent upon the decline of temperatures, a certain amount of retrograde metamorphism has followed prograde metamorphism, relics of the prograde minerals are frequently seen in the form of pseudomorphs composed of the retrograde minerals (see Fig. 10.6).

10.7 Garnet with trails of inclusions of elongate grains of quartz and iron ore. The overall arrangement of the trails gives a very shallow S-shape. × 75. Plane polarized light.

In addition to the above relict structures, which are seen in the general arrangement of the minerals in a metamorphic rock, we may recognize another group of relict structures which are preserved by the arrangement of inclusions within poikiloblasts. Thus examination of poikiloblastic rocks in thin section often reveals the inclusions within the poikiloblasts to be arranged in distinct lines or trails. The inclusions may be of many minerals but are most frequently of quartz, iron ore, and graphite, and their linear arrangement is a relic from some parallel structure which has been enveloped and preserved within the poikiloblast during growth. The original parallel structure may have been of either sedimentary (associated with bedding) or metamorphic (associated with rock cleavage) origin, and it may or may not be present in the matrix of the metamorphic rock. Sometimes the trails of inclusions in a poikiloblast are not

10.8 Garnet in mica-schist. Elongate quartz inclusions within the garnet have an S-shaped arrangement. × 75. Crossed nicols.

straight but curved. Such curvature may result from the relative rotation of the poikiloblast with respect to the matrix during growth, or alternatively from the enclosure of mineral grains which had of themselves a curved or folded arrangement. In the former case the inclusion trails usually show an S-shaped arrangement, symmetrical about the centre of the poikiloblast (see Figs. 10.7 and 10.8) while in the latter case (*helicitic structure*) the arrangement is not symmetrical.

Trails of inclusions within poikiloblasts are particularly common in regionally metamorphosed rocks, and in such cases are usually relics of metamorphic structures. It is sometimes possible to show by the comparison of such trails of inclusions with the cleavage structures present in the rocks, that not only several phases of deformation (see p. 208) but also several phases of mineral growth have occurred within the general period of movement and metamorphism associated with an orogeny. Further consideration of this subject is, however, beyond the scope of this work.

Naming a Rock. Metamorphic rock classification is attended by more difficulties than that of igneous and sedimentary rocks. This is largely a result of the much greater range of chemical composition of metamorphic rocks, and the fact that even one chemical composition may show many variations in texture, structure, and mineralogy depending upon the conditions of metamorphism. In addition it is always important to indicate the probable nature of a rock before metamorphism. Because of these features, the nomenclature of metamorphic rocks is not so systematic as that of the other two rock groups. However, we may roughly divide many of the names applied to contact and regional metamorphic rocks into two broad groups which are considered in turn below.

Terminology referring to the Nature of a Rock before Metamorphism. Many of these terms are simply derived by adding the prefix 'meta' to the name of the unmetamorphosed rock. Thus we may denote a metamorphosed igneous rock as being a *metagabbro*, a *metabasalt*, a *metarhyolite*, etc. Similarly in a broad way a metamorphosed sedimentary rock may be referred to as a *metasediment*, or more specifically as a *metagreywacke*, for example.

However, many of the names used more frequently to indicate the parentage of a metasediment do not follow the above principle. Instead, metamorphosed rudaceous, arenaceous, and argillaceous rocks are referred to respectively as *psephites*, *psammites*, and *pelites*. For metamorphosed limestones the nomenclature is not so straightforward. Metamorphosed limestones are themselves often referred to simply as '*limestones*', while other terms such as *marble* and *calc-silicate rock* are also employed. The latter names, however, have more specific mineralogical implications and, although it is convenient to note them here, they belong more strictly to the following section. A marble is a metamorphosed limestone which still retains a large proportion (generally over 50 per cent) of carbonates, while a calc-silicate rock consists predominantly of calcium-bearing silicates. A limestone containing considerable terrigenous impurity is usually transformed to a calc-silicate rock on metamorphism—

much of the carbonate being consumed in reactions with the impurities during the metamorphism.

Since the original mineralogy and texture of a rock are often destroyed during metamorphism, the application of names like pelite and metabasalt may not always be easy, especially where one is handling isolated hand-specimens in the laboratory. However, with approximately isochemical metamorphism, chemical composition, as indicated by mineralogy, always provides a valuable guide to the original nature of a rock (see Tables 8.1 and 9.1). It has been indicated that the mineralogy of a rock of a certain chemical composition will vary with grade, but we may still note some of the minerals which often occur in certain common contact and regional metamorphic rock types, even though these minerals will not all be formed under the same conditions of metamorphism. Thus *pelites* usually contain a small to moderate amount of quartz together with silicates rich in Al_2O_3 and/or FeO and MgO: muscovite, biotite, chlorite, cordierite*, garnet, staurolite, andalusite, kyanite, sillimanite, and hypersthene. The notable K_2O content of pelites is present in the micas and/or potash feldspar. A little plagioclase is usually present but it is always sodic, and minerals richer in CaO are scarce or absent. Since the cement in many sandstones is of an argillaceous nature we may consider many *psammites* as being chemically similar to the pelites but much diluted with quartz (and possibly feldspar). Thus quartz becomes the predominant mineral (generally over 70 per cent) in psammites, and it is usually accompanied by some mica and feldspar. Minerals such as garnet are not uncommon in small amounts but the aluminous minerals, such as kyanite, andalusite, sillimanite, staurolite, are much less frequently seen. For rocks intermediate between an obvious pelite and an obvious psammite the name *semi-pelite* is useful.

In basic igneous rocks, MgO, FeO, CaO, and Al_2O_3 are all major constituents and so *metabasalts*, *metadolerites*, and *metagabbros* consist of such minerals as: plagioclase, hornblende, epidote, chlorite, augite, hypersthene, garnet. Quartz and potash-bearing minerals, like biotite, muscovite, and orthoclase, are absent or present in only small amount.

Besides calcite and possibly dolomite, *marbles* may contain

* Cordierite is a silicate with the composition $(Mg,Fe)_2Al_4Si_5O_{18}$ and it frequently occurs in metamorphosed argillaceous sediments. However, it is a difficult mineral to identify and has therefore been omitted from the section on mineralogy.

silicates rich in CaO and/or MgO such as: diopside (a pyroxene similar to augite), grossular garnet, idocrase, calcic plagioclase, hornblende, epidote, tremolite, forsterite (olivine). In some marbles quartz is a prominent constituent. *Calc-silicate rocks* consist predominantly of combinations of some of the silicate minerals listed above for marbles, but with less or no carbonate.

The mineralogical distinctions between the above groups are therefore fairly clearcut, irrespective of changes due to grade. However, it should be noted that rocks of intermediate compositions and mineralogy blur these distinctions. Thus marls and other carbonate-bearing terrigenous sediments form metamorphic rocks whose mineralogy ranges from that of calc-silicate rocks to that of pelites and psammites.

Terminology indicating a Specific Structure, Texture, and/or Mineralogy. The following names, based upon structural features of contact and regionally metamorphosed rocks, are widely used:

schist—a rock with schistose structure
gneiss—a rock with gneissose structure
hornfels—a tough, massive rock.

Schists and gneisses are typical products of regional metamorphism, while hornfelses are essentially restricted to contact aureoles. Between them, these names cover many metamorphic rocks, and another term (*granulite*) has been used as a general name for rocks showing granulose structure. However, the term granulite has also been applied to indicate that a rock belongs to the granulite metamorphic facies (see later and Table 10.1), and its usage is best avoided in the structural sense.

Although the term schist is applicable to any schistose rock, fine-grained pelitic rocks with a well-developed schistosity of the type known as slaty cleavage (see p. 206) are usually termed *slates*. In the latter it is generally impossible to pick out individual mineral grains (excluding rare porphyroblasts) with the naked eye, whereas in those pelitic rocks commonly described as schists one can usually distinguish some individual grains of such minerals as micas and amphiboles. Schistose pelitic rocks with a grain size intermediate to these two categories are usually called *phyllites*, and such rocks often show a silky sheen on schistosity surfaces. Slates are typical products of very low grade regional metamorphism, and a sequence with increasing grade of slate to phyllite to pelitic schist is shown by the pelitic rocks in many regionally metamorphosed terrains.

Table 10.1

ESKOLA'S EIGHT METAMORPHIC FACIES

Facies	*General Conditions of Metamorphism*
GREENSCHIST FACIES	Many conditions of low-grade regional and contact metamorphism.
AMPHIBOLITE FACIES	Many conditions of middle- to high-grade regional and contact metamorphism.
ALBITE-EPIDOTE AMPHIBOLITE FACIES	Intermediate to those of the greenschist and amphibolite facies.
PYROXENE HORNFELS FACIES	Highest grades of metamorphism in many contact aureoles.
SANIDINITE FACIES	Exceptionally high-grade (and low-pressure) conditions of contact metamorphism. The relevant mineral assemblages are only seen immediately adjacent to igneous contacts and in xenoliths.
GRANULITE FACIES	Often high-grade regional metamorphism, but the characteristic mineral assemblages may also develop at lower temperatures in water-deficient environments.
GLAUCOPHANE SCHIST FACIES	Regional metamorphism where low temperatures have been combined with very high pressures.
ECLOGITE FACIES	This facies differs from all others in that its characteristic mineral assemblages are of basaltic bulk composition only. These mineral assemblages form at extremely high pressures in water-deficient environments.

In order to give information about the original nature, composition or mineralogy of a rock, adjectives are used in conjunction with the above purely structural rock terms, or compound hyphenated names are used. Thus we may refer to a pelitic or psammitic schist, a basic or calc-silicate hornfels, a feldspathic gneiss and so on. In the compound names constituent minerals are attached to the words thus giving, for example, an orthoclase-garnet-biotite-gneiss, a plagioclase-pyroxene-hornfels, or a hornblende-schist. In naming one particular rock the prefixed mineral names should normally be arranged in order of abundance, putting the most abundant first.

With regard to schists, however, it is the usual practice to depart from this rule to the extent of putting immediately before 'schist' the name of the inequidimensional mineral or minerals responsible for the schistosity. Obviously, to avoid names becoming too unwieldy, only the more important mineral constituents are indicated in the rock name. Thus a plagioclase-pyroxene-gneiss is a rock with gneissose structure, and whose principal constituents are plagioclase and pyroxene, the plagioclase being present in greater amount than the pyroxene. Similarly a quartz-garnet-mica-schist is a schistose rock whose schistosity is a result of the parallel orientation of mica flakes (usually biotite and muscovite) and in which quartz and garnet are also important constituents, the former being more abundant than the latter.

In addition to the above generally applicable compound names, other terms are used in specific cases in order to indicate a certain mineralogy or origin together with a certain texture or structure. Thus a *quartzite* is a tough and massive rock consisting almost wholly of quartz, and is the usual product of contact or regional metamorphism of an orthoquartzite.* For a psammite of similar structure to a quartzite but including a fair amount of feldspar besides quartz, the name *feldspathic quartzite* may be used.

Another rock type whose distinction rests upon a combination of criteria is a *migmatite*. Such a rock is often coarsely gneissose with some bands and patches consisting largely of quartz and feldspar with a granitic texture, while other bands are relatively rich in biotite or hornblende and have a crystalloblastic appearance. Migmatites thus have the aspect of being formed of a mixture of igneous and metamorphic elements. They are commonly seen in areas of high-grade regional metamorphism.

Finally, certain names relate solely to the mineralogy of a metamorphic rock. The terms *marble* and *calc-silicate rock*, recognized in the previous section, strictly belong here. In addition we may note the name *amphibolite*, which is frequently used to refer to a rock composed largely of hornblende and plagioclase irrespective of its texture or structure. Compound names based on such terms as amphibolite and marble are also useful. Thus many amphibolites contain garnet and such rocks are referred to as garnet-amphi-

* Under exceptionally high shearing stress an orthoquartzite may acquire a marked directional structure because of extreme flattening or elongation of the quartz grains, and then the name quartz-schist becomes more appropriate.

bolites. Similarly, in order to indicate which important minerals, besides carbonate, a marble contains, we may use such names as forsterite-marble, diopside-grossular-marble, etc.

Amphibolites may result from the metamorphism of igneous (basaltic) or sedimentary (carbonate-bearing shale) rocks and it is often very difficult to distinguish between these two origins. Where the parentage can be determined then an amphibolite may be referred to as an ortho-amphibolite if of igneous origin or a para-amphibolite if of sedimentary origin. The prefixes 'ortho' and 'para' are often similarly used to indicate the parentage of gneisses.

DISLOCATION METAMORPHISM

Some features of terminology and its usage as discussed above for the structures and nomenclature of contact and regionally metamorphosed rocks, also apply to the products of dislocation metamorphism. However, the common combination of high shearing stress and low temperature, resulting in considerable deformation and little mineral growth during dislocation metamorphism, gives rise to certain structures and rock types which have no counterparts in contact and regional metamorphism. Also as a result of the limited amount of recrystallization and neomineralization occurring during dislocation metamorphism, mechanical features of pre-existing rocks become much more important than chemical features.

In the case of such soft, fine-grained rocks as argillaceous sediments, the deformation accompanying dislocation metamorphism is accommodated by continuous movement and adjustment throughout the body of rock. Coupled with limited mineral growth this leads to the formation of a rock with the features of a *slate*, and to the naked eye the rock appears to have yielded to the deformation as a whole and shows no planes or zones of fracture.

In contradistinction to the above are the much harder and more brittle rocks, such as igneous rocks and high-grade metamorphic rocks, for at low temperatures these are not capable of yielding in the same manner as the softer and less brittle rocks. Instead, they deform in a more discontinuous manner, and lines and zones of fracture are often conspicuous to the naked eye. Crushing and granulation of minerals (*cataclasis*) occurs within the fracture zones, and depending upon the intensity of deformation a large variety of cataclastic structures and rocks may result, some of which are detailed below.

Crush-breccia—a rock consisting of angular fragments (visible to the naked eye) of the original rock.

Kakirite—similar to a crush-breccia but in which the angular fragments are surrounded by fine-grained material produced by crushing of part of the original rock.

Flaser rocks—those having lenticles (visible to the naked eye) of relatively unaltered rock-fragments or minerals (termed *porphyroclasts*) enclosed in a highly crushed and granulated matrix, the latter usually having partially recrystallized. We may distinguish such rocks as being a flaser-gabbro, flaser-granite, flaser-gneiss, etc., according to the nature of the original rock. Flaser rocks result from greater shearing or differential movement within a rock than occurs in the formation of a crush-breccia or kakirite.

Augen-gneiss—a flaser rock in which the lenticles (augen) consist of single crystals of an original mineral (usually feldspar).

Mortar structure—that structure in which relatively large relics of original minerals are embedded in a finely crushed matrix.

Cataclasite—a rock consisting of finely ground, structureless rock powder with few or no porphyroclasts. Such rocks indicate a further stage of crushing and granulation than the flaser rocks.

Mylonite—another rock formed by extreme crushing and granulation but differing from a cataclasite in showing a banded structure and often a platy fracture. Mylonites present the appearance of having been powerfully rolled out or milled by extreme differential movement.

EXAMINATION AND DESCRIPTION OF A METAMORPHIC ROCK

The microscopic examination of metamorphic rocks is generally more difficult than that of igneous rocks, because the range of mineralogy is greater, and some of the minerals are by no means easy to determine. Particular difficulty is found with the distinction of minerals such as potash feldspar, plagioclase, and quartz because of the common lack of twinning in the feldspars. Material with the general appearance of these minerals should be examined carefully with the light stopped down by means of the substage diaphragm. By this means the student should be able to pick up small differences in relief and so ascertain whether aggregates of such material consist of one or several minerals. Also the cleavage of feldspars becomes clearer under such illumination. Refractive indices of the mineral grains should be compared with that of Canada Balsam. Traces of twinning in plagioclase or microcline will often indicate the presence of these minerals, even though the twinning is usually not sufficiently well developed to provide a means of estimating the proportions of the minerals. In many rock types, such as pelites and metabasalts, potash feldspar only occurs when they have been subjected to high-grade metamorphism, and in these cases the potash feldspar is orthoclase and may show fine perthitic structure. Finally, feldspars may show slight retrograde alteration (e.g., cloudiness, the presence of sericite), which will help in their distinction from quartz.

In writing out a description of a metamorphic rock, observations should be presented in an orderly and logical manner, and the following general procedure is recommended.

(a) Description of hand-specimen, e.g., colour, structure, and texture of the rock, relation of cleavage to bedding, minerals identifiable in hand-specimen.

(b) Description of thin section.

(i) Brief and comprehensive description of the mineralogy, and main features of texture and structure of the rock. This should include a list of the minerals present in their order of abundance.

(ii) Description of individual minerals and more detailed account of textures and structures where necessary. Descriptions of minerals should include only inherently variable features (see p. 166). The composition of plagioclase, if present, should be noted where possible. As a general rule it is best to describe the minerals in their order of abundance, noting their approximate volume proportions as far as they may be ascertained.

However, in the case of poikiloblastic and porphyroblastic rocks the matrix minerals should usually be described first together with any texture or structure (e.g., schistosity) their arrangement gives rise to. This should be followed by description of the poikiloblastic and/or porphyroblastic minerals, noting their relation to such structures as schistosity if present (e.g., schistosity curves around porphyroblasts) and the nature of any inclusions they contain. The grain size of included minerals should be compared with that of the same minerals in the matrix. In addition the arrangement of the inclusions (e.g., forming straight trails; forming S-shaped trails) should be noted, and also the relation of inclusion trails to any parallel arrangement of minerals in the matrix (e.g., straight trails parallel to schistosity; S-shaped trails arranged so that the ends of the 'S' are parallel to the schistosity).

(iii) Conclusions made on the basis of described observations. A name should be given to the rock, and should be set at the top of the description together with the locality. It is clear, from the discussion in preceding pages, that a large number of names may be applied to a single rock. Thus a schistose rock consisting largely of biotite and muscovite with garnet and a little quartz and plagioclase, might be termed a metasediment, a pelite, a pelitic schist, or a garnet-mica-schist. From the viewpoint of general description, however, the student should always apply that name which conveys most information about a rock. For contact and regionally metamorphosed rocks, this name will usually be of the compound type such as garnet-mica-schist, plagioclase-hornblende-schist, diopside-marble, etc. Such names are also preferable, when describing isolated specimens in the laboratory, because they are based on directly observable features of the rock.

Inferences as to the probable original nature of the rock, and the kind of metamorphism (i.e., contact, regional, or dislocation) that has produced it should be made after applying a name as above. In some cases, however, the actual name applied (e.g., flaser-granite) will have already involved consideration of these factors. Inferences as to the probable or possible original nature of the rock before metamorphism will be based upon relict structures, and the minerals present and their proportions. The common differences between the structures formed under contact, regional and dislocation metamorphism, will usually allow the student to suggest which kind of metamorphism has produced a particular rock. Difficulties can arise in some cases, however, e.g., marbles typically show a granulose structure whether formed by contact or regional metamorphism.

Metamorphic Facies and Classification of Types of Regional Metamorphism

No attempt has been made, in the preceding pages, to cover such important genetic aspects of metamorphic rocks as the estimation and comparison of the temperatures and pressures at which various rocks are formed. Certain classifications of metamorphic rocks have much significance in this theoretical context, and their nature is outlined in the following pages.

On p. 202 we noted that changes of grade are often indicated by changes of mineral assemblages in rocks of given chemical composition. Such variation in mineral assemblages is common within areas of both contact and regional metamorphism, and the various assemblages in any one area are usually arranged in regular zones, termed *metamorphic zones*. The use of such zones therefore provides a means of classifying the rocks of a particular area with regard to variation in the conditions of metamorphism. Zonal classifications, however, do not provide a general scheme for the comparative study of rocks of all metamorphic terrains. Another classification, in which rocks are placed in *metamorphic facies*, is much better suited to this objective, and also provides a background against which different sequences of metamorphic zones may be compared. To cover the metamorphic facies classification in sufficient detail to enable the student to classify a rock within it, is beyond the practical scope of this book, because advanced mineralogical and chemical techniques are often necessary. Because of the usefulness of the classification, however, the student should at least be aware of its nature and objectives, besides which some knowledge of the classification adds significance to the careful examination and description of metamorphic rocks.

In the *metamorphic facies classification*, metamorphic rocks are put into groups (facies) on the basis of the relationship between their chemical composition and their mineralogy. Ideally, each rock composition should be represented in each facies by a particular mineral assemblage, and each facies should consist of a group of mineral assemblages (all representing different ranges of bulk chemical composition) formed under similar conditions of metamorphism. We may perhaps explain this more clearly by considering a hypothetical example. Let us suppose that a geologist is working in an area largely composed of thin flows of basalt lava altered by either contact or regional metamorphism. At one outcrop in this area he finds metabasalts characterized by a number of mineral assemblages, which differ slightly from one another because of small differences in chemical composition between the original basalt flows. These mineral assemblages, therefore, belong to one metamorphic facies. Some distance away, at another outcrop, the geologist finds a group of metabasalts with the same range of chemical compositions as those at the first outcrop, but he finds that the mineral assemblages in corresponding flows are different. This difference in mineral assemblages between the two outcrops, despite the fact that both outcrops show the same range of bulk chemical compositions, means that the conditions of metamorphism must have been different at the two localities. The mineral assemblages at the second outcrop, therefore, belong to a different facies from that of the mineral assemblages of the first outcrop.

The above example illustrates how the geologist decides which mineral assemblages belong to the same and which to different, metamorphic facies. *Rocks from different metamorphic terrains are assigned to the same facies if they show the same mineralogy for the same chemical composition.* Close association of various rock types in the field is always important, because it provides a means of determining which mineral assemblages (of differing bulk chemical composition) form under the same conditions of metamorphism. For example, by noting the mineral assemblages developed in pelitic rocks at field locations at which metabasalts also occur, it is possible to determine which mineral assemblages in pelitic rocks form under the same conditions of metamorphism (and therefore belong to the same facies) as certain mineral assemblages in metabasalts. Therefore, each facies is defined by the field association of a *group* of mineral assemblages developed in appropriate bulk chemical compositions.

Confining ourselves to prograde metamorphism and to those conditions of water content, etc., indicated on p. 200, it is clear that the differences in conditions of metamorphism, which each facies represents, are differences in temperature and pressure, providing we also assume that the mineral assemblages developed in the rocks approximate to chemical equilibrium.

Since the precise temperatures and pressures at which one rock chemical composition (e.g., a metabasalt) shows a change of mineral assemblage will generally be different from those at which a rock of markedly different chemical composition (e.g., a pelite) shows a change of assemblage, it is ideally possible to define an exceptionally large number of facies. To do this, however, would destroy many of the advantages of the facies classification as a general scheme useful to all geologists. Consequently, geologists prefer to recognize only a small number of facies, each of which corresponds to large ranges of temperature and pressure.

The metamorphic facies commonly encountered are eight in number, and they were first proposed by Eskola. They are based to a large degree upon the mineral assemblages found in metamorphosed basic igneous rocks of normal basaltic composition (hence the choice of metabasalts in the earlier example). However, it must be noted that the use of only a small number of facies has important consequences. Thus, certain bulk chemical compositions (e.g., pelitic compositions), which are particularly sensitive to changes in temperature and pressure, show several different mineral assemblages within the range of conditions encompassed by one facies. Conversely, other chemical compositions (e.g., that of a quartzite), which are largely insensitive to changes in temperature and pressure, are represented by the same mineral assemblage in many facies. It must be emphasized, therefore, that the small number of facies recognized can only be completely defined in terms of a large number of mineral assemblages developed in rocks of widely different chemical compositions.

Eskola's eight facies are listed in Table 10.1 (see p. 216) together with the metamorphic conditions (largely in terms of grade) to which they correspond. The names given to the facies are based partly upon the mineral assemblages found in metamorphosed basic igneous rocks, but the names are only general designations and should not be interpreted too literally. Estimated temperature and pressure conditions corresponding to each facies are shown in Fig. 10.9. The boundary lines drawn between the facies on this diagram are by no means certain, and it is one of the objectives of metamorphic petrology to define them as accurately as possible.

The value of the combined practical and theoretical nature of the metamorphic facies classification has been clearly demonstrated in recent years by the classification of various types of regional metamorphism. At the end of the nineteenth century, Barrow, working in a regionally metamorphosed terrain in the south-eastern Highlands of Scotland, recognized a sequence of metamorphic zones based upon minerals (e.g., garnet, staurolite, kyanite, sillimanite) found in pelitic rocks. The mineral assemblages developed in these rocks belong to the greenschist, albite-epidote amphibolite, and amphi-

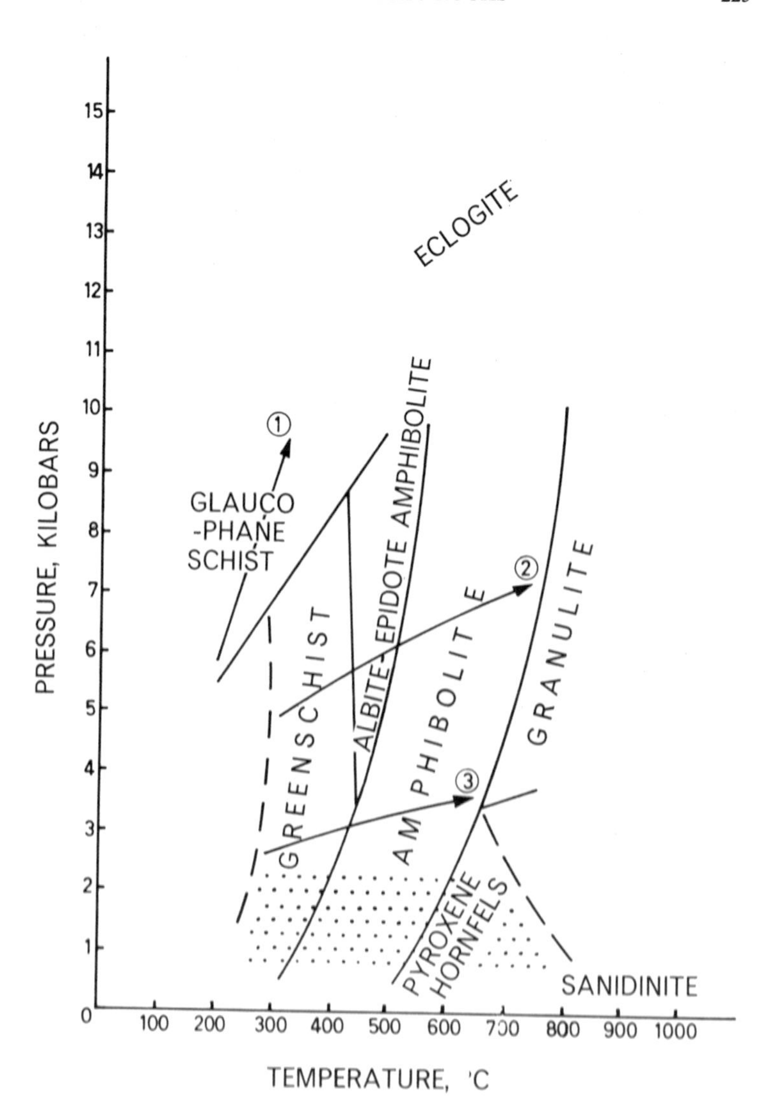

10.9 Estimated temperature and pressure ranges corresponding to Eskola's eight metamorphic facies. The arrowed lines indicate typical temperature/pressure curves for the three standard types of regional metamorphism: (1) Glaucophane-schist facies type; (2) Barrovian type; (3) Abukuma type. Temperatures and pressures commonly realized in contact metamorphism are shown by the dotted region.

bolite facies; differences between the zones (six in total) being largely the result of regular variation in temperature at the time of metamorphism. For many years it was considered that regional metamorphism normally led to the formation of pelitic mineral assemblages of Barrovian type. It is now apparent that pelitic mineral assemblages may vary largely as a result of differences in pressure between different areas of regional metamorphism. Thus, Miyashiro has recognized three standard types of regional metamorphism between which various intermediates occur. The standard types* and their relative pressure conditions are:

(a) Glaucophane-schist facies type—very high pressure
(b) Barrovian type —high pressure
(c) Abukuma type —low pressure.

Typical temperature/pressure curves for the three standard types of regional metamorphism are shown in Fig. 10.9. Also indicated, in a belt running across the diagram, are the temperatures and pressures commonly developed during contact metamorphism. The approach of the pressure and temperature conditions of Abukuma type regional metamorphism to those of common contact metamorphism, is demonstrated by the similarity of the pelitic mineral assemblages developed under the two sets of conditions. In both cases pelitic rocks usually contain such minerals as cordierite (see p. 214), andalusite, and sillimanite, while typical minerals (e.g., kyanite, staurolite) of Barrovian metamorphism are absent.

* Alternative names may be used for the standard types, those used here being chosen for simplicity. Abukuma is the name of a plateau in Japan.

Index

1234567890 WC 7432106987

THIS BOOK HAS BEEN SET IN MONOPHOTO TIMES NEW ROMAN 10 ON 12 POINT
AND PRINTED AND BOUND IN GREAT BRITAIN BY
WILLIAM CLOWES AND SONS, LIMITED, LONDON AND BECCLES